Modeling Dynamic Systems

Series Editors

Matthias Ruth
Bruce Hannon

Springer
New York
Berlin
Heidelberg
Barcelona
Hong Kong
London
Milan
Paris
Singapore
Tokyo

MODELING DYNAMIC SYSTEMS

Modeling Dynamic Biological Systems
Bruce Hannon and Matthias Ruth

Modeling Dynamic Economic Systems
Matthias Ruth and Bruce Hannon

Dynamic Modeling in the Health Sciences
James L. Hargrove

Modeling and Simulation in Science and
Mathematics Education
Wallace Feurzeig and Nancy Roberts

Dynamic Modeling of Environmental Systems
Michael L. Deaton and James J. Winebrake

Dynamic Modeling, Second Edition
Bruce Hannon and Matthias Ruth

Modeling Dynamic Climate Systems
Walter A. Robinson

For more information, see:
www.springer-ny.com/biology/moddysys

Walter A. Robinson

Modeling Dynamic Climate Systems

With 169 Illustrations

 Includes CD-ROM

Springer

Walter A. Robinson
Department of Atmospheric Sciences
University of Illinois at Urbana-
  Champaign
105 South Gregory Street
Urbana, IL 61801, USA
robinson@atmos.uiuc.edu

Series Editors:

Matthias Ruth
Environment Program
School of Public Affairs
University of Maryland
3139 Van Munching Hall
College Park, MD 20742-1821, USA

Bruce Hannon
Department of Geography
220 Davenport Hall, MC 150
University of Illinois
Urbana, IL 61801, USA

Cover photograph: Clouds over the Pacific Ocean as seen in a false color enhanced infrared image from the GOES-10 satellite. Courtesy Department of Atmospheric Sciences, University of Illinois at Urbana-Champaign, ww2010.atmos.uiuc.edu.

Library of Congress Cataloging-in-Publication Data
Robinson, Walter A., 1955–
    Modeling dynamic climate systems / Walter A. Robinson.
        p.   cm. — (Modeling dynamic systems)
    Includes bibliographical references and index.
    ISBN 0-387-95134-2 (alk. paper)
        1. Climatology—Mathematical models.   2. Atmospheric physics—Mathematical models.
    3. Ocean-atmosphere interaction—Mathematical models.   I. Title.   II. Series.
QC981.R63 2000
551.6′01′5118–dc21   00-061864

Printed on acid-free paper.

Production coordinated by Impressions Book and Journal Services, Inc., and managed by Timothy A. Taylor; manufacturing supervised by Jerome Basma.
Typeset by Impressions Book and Journal Services, Inc., Madison, WI.
Printed and bound by Maple-Vail Book Manufacturing Group, York, PA.
Printed in the United States of America.

9 8 7 6 5 4 3 2 1

ISBN 0-387-95134-2   SPIN 10779910

Springer-Verlag   New York   Berlin   Heidelberg
A member of BertelsmannSpringer Science+Business Media GmbH

Series Preface

The world consists of many complex systems, ranging from our own bodies to ecosystems to economic systems. Despite their diversity, complex systems have many structural and functional features in common that can be effectively modeled using powerful, user-friendly software. As a result, virtually anyone can explore the nature of complex systems and their dynamical behavior under a range of assumptions and conditions. This ability to model dynamic systems is already having a powerful influence on teaching and studying complexity.

The books in this series will promote this revolution in "systems thinking" by integrating skills of numeracy and techniques of dynamic modeling into a variety of disciplines. The unifying theme across the series will be the power and simplicity of the model-building process, and all books are designed to engage readers in developing their own models for exploration of the dynamics of systems that are of interest to them.

Modeling Dynamic Systems does not endorse any particular modeling paradigm or software. Rather, the volumes in the series will emphasize simplicity of learning, expressive power, and the speed of execution as priorities that will facilitate deeper system understanding.

Matthias Ruth and Bruce Hannon

Preface

When I was a teenager, my father taught me to do some minor sorts of maintenance and repair work on our family car—a Dodge Dart. I was never good at this—too clumsy—but from looking around under the hood, I started to understand how a car works: how the various components—cylinders, ignition, carburetor, and so on—interact as a system to make the car run. In comparison with a year-2000 automobile, the Dart was good to work on and good for learning about cars. It was simple. All the key parts were plainly visible, and their functions were obvious. One look at the breaker points opening and closing, and I could tell that they were a kind of switch. The smell of gasoline from the carburetor meant it had to do with getting fuel to the engine, and since it was right inside the air filter, air as well. And the Dart was tough. When I made a mistake, like connecting the wires from the distributor to the wrong spark plugs, I did not wreck the engine. In late-model cars, by contrast, the parts are concealed, electronic, and unserviceable by anyone other than a trained mechanic.

What does this have to do with climate? Modern climate models, the global climate models (GCMs) used to make projections of global climate change, are big, complicated, and expensive packages of computer code that make huge demands on the best computer hardware. A student or layperson rarely gets a chance to work with such a model, and, even when one does, it has to be treated for the most part like a "black box." Only trained programmers get to make changes in the code.

By comparison, the models described in this book are simple. They have the advantage, not unlike a Dodge Dart, that to a great extent how they work is self-evident. And, since they are implemented in STELLA, they can be "broken" and "fixed" without a huge investment in programming or computing time. Most people would not care to drive a Dart today. Modern vehicles are safer, more efficient, less polluting, and more reliable. Similarly, one would not want to use the models in this book for making quantitative predictions of climatic change. What these models *are* good for is to develop an understanding of how the climate works, first by repeating the baseline runs and experiments described here, then by doing the suggested problems, and finally and most importantly, experimenting with modifications of one's own invention. The reader should be fearless in modifying the models. If one does not break them, one is probably not learning as much as one can.

The attraction for me, and I hope for the reader, of systems modeling with STELLA, is that it is fun. Models are quickly constructed and even more quickly modified. One knows in minutes whether a model produces interesting and, when plotted, esthetically appealing behavior, produces a boring straight line, or blows up into numerical nonsense. So, my fondest hope for this book is that it will make enjoyable the reader's study of climate dynamics.

Two cautions to the reader. First, to model climatic processes in the context of STELLA, many simplifications of complex processes must be made. Many arbitrary choices have been made in constructing these models, and I make no claim that my choices are unique, or even the best. Readers will know they are learning from this book if they come up with better ways to build these models or can point out errors in my versions. Of course, I would appreciate hearing the results of such improvements; my postal and electronic addresses are given below. In defense of these simple models, it should be pointed out that GCMs also include arbitrary choices and simplifications of complex processes (these are called *parameterizations*), but they are often hidden deep inside complex model codes. Working with STELLA models thus provides an excellent introduction to any level of the modeling process.

Second, while brief explanations of the processes working in each model are given here, this book is about learning by doing and does not pretend to be a comprehensive textbook on the physics of climate. Excellent texts of this type are available; my favorite is *Global Physical Climatology*, by Professor Dennis Hartmann (1994, Academic Press, 411 pp.). The reader is encouraged to work through the models here with a book of this sort close at hand.

The book is organized as follows: Chapter 1 introduces STELLA modeling in the context of two very simple models. It also introduces two paradigms for stability. The first of these, a particle in a well, is relevant to the dynamics of individual parts of the climate system, such as a parcel of air in a stable sounding, but it is the second, the "leaky bucket," that is more relevant to the climate as a whole. Chapter 2 introduces a number of climate models based on energy balance. The first are globally averaged. Later in the chapter are models that resolve the climate, if coarsely, along the spatial dimensions of height or latitude. Chapters 3 through 5 deal with the dynamics of individual parts of the climate system: The vertical motions of parcels of air are treated in Chapter 3, their horizontal motions are treated in Chapter 4, while Chapter 5 deals with the dynamics of the more abstract quantities of circulation and vorticity. Chapter 6 then returns to models of climate but models that are more dynamic, in that they explicitly include some representation of atmospheric flow. Finally, Chapter 7 presents models that represent different aspects of climate variability, and it returns to the ideas of stability and sensitivity first introduced in Chapter 1.

Contents

Acknowledgments

The idea for this book and the suggestion that I write it came from an editor of this series, Professor Bruce Hannon, who is a colleague at the University of Illinois. Professor Hannon has been spreading the gospel of STELLA, systems thinking, and systems modeling for many years, and he has helped bring a much-needed environmental consciousness to our campus. Some of the models here were developed as student exercises for a course I co-teach at Illinois, now titled Modeling Earth and Environmental Systems. I am grateful to the students in this course for showing me what aspects of the models and modeling are hard to "get" and for pointing out some errors in earlier versions of these models. My co-teacher, Professor Scott Isard, brought a broader view of systems thinking and modeling to the course than I encountered in my own physics-based education and research. I hope some of his approach has rubbed off on this book.

Last, but certainly not least, I acknowledge my father, Dan Robinson, who not only showed me how to work on a Dodge Dart but also led me, as far back as I can remember, to the conviction that things happen in the world for reasons that can be rationally analyzed and understood. It is to him, with love, that this book is dedicated.

Walter A. Robinson

1

Two Paradigms of Stability

1.1 Stability and Instability

Life has existed on Earth for more than 3 billion years. This probably requires the presence somewhere on the planet of temperatures between the freezing and boiling points of water (0 to 100°C). That the climate has remained within such narrow bounds for so long strongly suggests that it is stable. It is expected intuitively that any long-lived system is stable in some way. This is why "balancing" rocks are so surprising. But what exactly do we mean when we say that a system is stable or unstable? The typical picture, found in many textbooks, is that of a ball in a valley or on top of a hill (Figure 1.1). This is a simple system to model, especially if we neglect the rotation of the ball and imagine that it slides without friction down the slope. Furthermore, the model of this system provides a good working introduction to STELLA.

In STELLA, the fundamental quantities that define the state of any system are identified as *stocks*. A stock is a quantity of some stuff, be it water, heat, momentum, or marshmallows. The quantities that change the values of the stocks are identified as *flows*. Flows have the units of the stock into or out of which they flow, divided by time. For the particle in the well, the fundamental quantities are the position and the velocity of the particle. Position changes with time according to the velocity, so the velocity must be a flow, into or out of the stock, Position, as well as a stock (note that STELLA requires that they be given different names). Velocity changes with time according to the acceleration, so the acceleration is a flow into and out of the stock, Velocity.

By Newton's second law of motion, the acceleration is the force divided by the mass. Here, the force along the path of the ball is the force of gravity along the slope of the surface. Now we must define how this force depends on the position of the ball. This depends on the shape of our hill or valley, but we can make the simplest assumption, that the force increases or decreases in proportion to the displacement of the ball from its initial position on the top of the hill or in the middle of the valley. Technically, this model is valid only for a frictionless particle of unit mass sliding in a parabolic

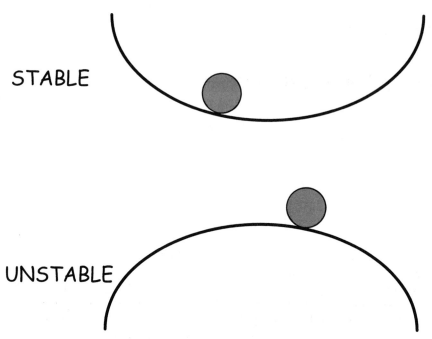

FIGURE 1.1   The simplest model of stability and instability.

valley or on a parabolic hill and only for small displacements of the particle. The acceleration of the particle is given by

$$\text{Acceleration} = \text{Constant} \times \text{Position}.$$

For a valley, the acceleration is to the right if the ball is displaced to the left, and vice versa, so the constant is negative. This is the simplest example of a system with negative feedback. For a hill, however, the acceleration is to the left if the ball is displaced to the left, and vice versa, so the constant is positive. This is the simplest example of a system with positive feedback. The model is shown in Figure 1.2. Note that the position, velocity, and acceleration can all take on negative as well as positive values, so the stocks must not be designated "Non-negative" in their dialogue boxes, and "BIFLOW" must be selected for the flows. This model is run, starting with the ball centered (Position = 0); the ball is given a small initial positive velocity, and the value of the stock, Position, is plotted against time. The result for the valley is a sinusoidal oscillation (Figure 1.3) but an increasing exponential for the hill (Figure 1.4). If Constant is set to zero, the model represents a flat surface, and, since there is no friction in this model, the ball maintains a constant velocity forever. This is the case of neutral stability—there is no positive or negative feedback—and Position increases or decreases linearly with time.

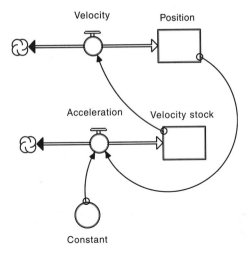

FIGURE 1.2   A STELLA model of a ball in a valley or on a hill.

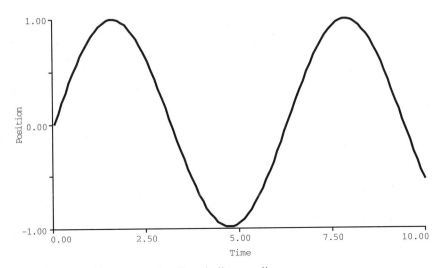

FIGURE 1.3   Position versus time for a ball in a valley.

This system is the standard textbook example of stability or instability, and it is a good starting point for learning STELLA, but is it relevant to the atmosphere or the climate? In Chapters 3 and 4 we will see that this type of stability or instability is similar to the mechanical behavior of a parcel of air displaced vertically or horizontally in the atmosphere. The ball in a valley is, however, a poor metaphor for the stability of the climate, because it is a completely closed system. No energy flows in or out, so the total energy of the system, the sum of the kinetic energy and the gravitational potential

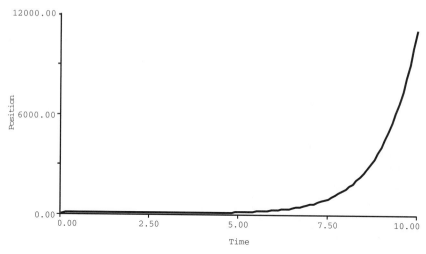

FIGURE 1.4 Position versus time for a ball on a hill.

energy of the ball, does not change with time. By contrast, the climate system is open with respect to energy. Energy arrives in the form of sunlight and is lost to space in the form of infrared radiation. The next model, the leaky bucket, is a very simple open system, and it is a far better metaphor for the climate.

1.2 The Leaky Bucket

Consider a bucket placed under a running tap. If the bucket is intact, it will eventually fill up and overflow. If there is a hole inside of the bucket near its bottom, water will run out. If the hole is big enough, the system may come to a state of equilibrium in which water runs out through the hole at the same rate that it flows into the bucket from the tap. Later, when we construct models of the global climate, we will see that this is analogous to the climate system. In this analogy, the volume of water in the bucket corresponds to the amount of thermal energy in the climate system (which is related to the average temperature of the earth), the flow of water from the tap corresponds to the flux of solar energy reaching the earth, and the water escaping through the hole corresponds to the thermal radiation that escapes from the earth into space.

To construct a model of the leaky bucket, it is necessary to recognize, first, that the flow of water from the tap does not depend on anything that happens inside the bucket. The rate at which water runs out the hole, however, depends on the pressure of water at the hole, and this in turn depends on the depth of water in the bucket. This dependence is given by Torricelli's theorem. The speed at which water leaves the bucket is given by

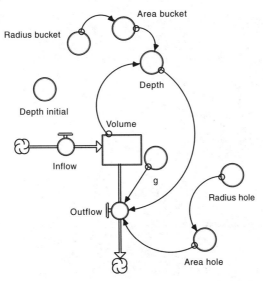

FIGURE 1.5   A STELLA model of the leaky bucket.

$$Speed = SQRT (2 \times g \times Depth),$$

where g is the acceleration of gravity (9.81 m s^{-2}) and Depth is the vertical distance between the center of the hole and the surface of the water in the bucket. The rate at which water leaves the bucket is given by its speed of flow times the area of the hole,

$$Outflow = Speed \times PI \times Radius\_hole^2.$$

The greater the depth of water in the bucket, the faster it runs out through the hole. This is a negative feedback for this system, in that a higher water level leads to a faster loss of water through the hole, which tends to lower the water level.

The heart of a STELLA model of this system (Figure 1.5) is a reservoir representing the volume of water in the bucket. Assume that the bucket is perfectly cylindrical, and, for, simplicity, assume that the hole is at the bottom of the bucket. Then the volume of water in the bucket is related to its depth, by

$$Volume = Depth \times PI \times Radius\_bucket^2.$$

In our model, Depth is a converter that depends on Volume according to

$$Depth = Volume/[PI \times Radius\_bucket^2].$$

It is easiest to think about this system if physically reasonable dimensions are used, say 0.5 m for the radius of the bucket, and 0.5 cm (0.005 m) for the radius of the hole. The loss of water, in units of cubic meters per second

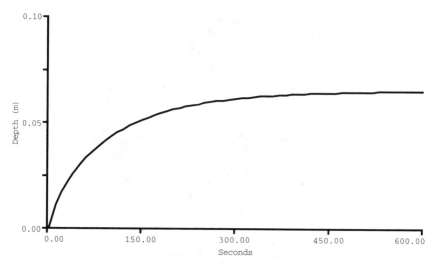

FIGURE 1.6   The depth of the water in the leaky bucket.

(these are small numbers) is given by the above relations. It is necessary to specify a rate, also in cubic meters per second, at which water fills the bucket from the tap.

When the model is run starting from an empty bucket, the bucket initially fills rapidly. Water is coming in, but, because the water level is still low, little is going out. As the bucket fills, the rate of flow out through the hole increases until it equals the rate at which water is coming in from the tap, and eventually an equilibrium is reached (Figure 1.6). Plotting the rates of inflow and outflow on the same graph (Figure 1.7) verifies that the rates of inflow and outflow are equal once an equilibrium depth is achieved. Better still, this behavior can be observed by punching a hole near the bottom of a tin can and putting it under the tap in your kitchen sink.

When the ball in the valley is in its equilibrium state, it is standing still; nothing is happening. When the water level in the bucket has stopped changing, there is still plenty happening. Water pours into the top of the bucket and rushes out through the hole in the bottom. Such dynamical equilibria are ubiquitous in nature. While the temperature of the earth is relatively steady, energy constantly arrives at the earth as sunlight and constantly streams into space as infrared radiation. When a glass of water sits in a sealed chamber, though the amount of water in the glass does not change, countless water molecules constantly leave the surface of the water and enter the vapor phase while an equally large number of vapor molecules arrive at and stick to the surface of the water.

Such dynamical equilibria are also fundamental for understanding many issues in the global environment. For example, it is often believed that the problem of stratospheric ozone depletion—the ozone "hole"—involves a onetime removal of ozone from the atmosphere. On the contrary, the key

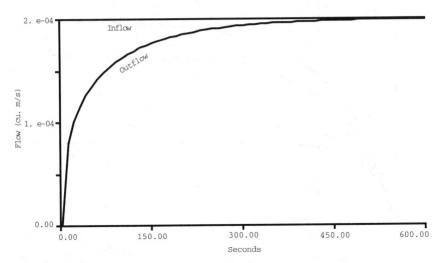

FIGURE 1.7   Flows of water into and out of the leaky bucket.

to understanding ozone depletion is recognizing that chemicals released into the air by humans increase the rate of ozone destruction, but that this rate also depends on how much ozone is present. At the same time, ozone is constantly being created. The net effect of ozone-depleting chemicals is that the ozone layer comes to a new equilibrium with a concentration of ozone lower than before. This is perfectly analogous to making the hole in the bucket larger. The water level drops to a new, lower equilibrium.

Before leaving the leaky bucket for systems that are more obviously related to climate, it can be used to demonstrate some basic concepts. First, we need to be convinced that the equilibrium water level in the bucket is truly stable. This can be done by running the model to equilibrium, noting the final value of the water level, and then starting a new run with an initial water level either a little higher or a little lower than this equilibrium value. The water level rapidly returns to its equilibrium. Note that, unlike the ball in a valley, this system does not overshoot its equilibrium or oscillate about it.

The leaky bucket also exhibits a property that is of critical importance for the climate—sensitivity to changes in external parameters. For the climate, we care a great deal how much warmer it will get when the concentration of carbon dioxide in the atmosphere is increased by a certain amount. For the leaky bucket, we can ask the more mundane question, "How much will the water level rise if we open the tap?" Such questions are readily addressed by using the "Sensi Specs" feature of STELLA to carry out model runs with different rates of inflow. Here, the sensitivity is how much the equilibrium water level changes for a given change in inflow.

The sensitivity of a system to changes in external forcing is itself sensitive to the values of its parameters. In the case of the leaky bucket, the sensitivity depends on the size of the hole. We can see this by using "Sensi Specs" under the "Run" menu to specify the parameters for four model runs with

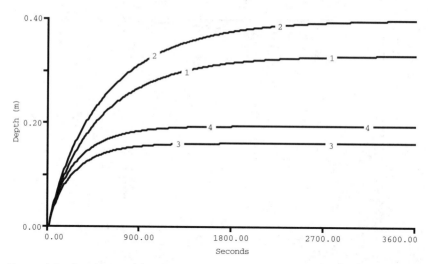

FIGURE 1.8   Sensitivity of the depth to a 10% increase in the inflow for holes with a radius of 5 mm (curves 1 and 2) and 10 mm (curves 3 and 4).

two different values of Inflow and two different values of Radius_hole. The results of all four runs are plotted on the same graph by selecting the "Comparative" feature in the graph dialogue box (Figure 1.8). The change in the equilibrium water level for a 10% increase in Inflow is greater when the hole has a radius of 5 mm (curves 1 and 2) than when its radius is increased to 6 mm (curves 3 and 4). Looking more closely at these curves, it is seen that the equilibrium water level is reached in less time when the hole is larger. Thus, at least for the leaky bucket, higher sensitivity corresponds to a slower approach to equilibrium or a slower return to equilibrium when the system is perturbed. That this is also true for earth's climate is of considerable practical importance. We return to this point in Chapter 7.

Most systems have some noise in them. Sometimes this comes from outside of the system, and sometimes it is produced internally. Certainly, there is noise in the climate system: external noise from variations in the output of the sun, and internal noise from year-to-year fluctuations in the exchanges of heat between the atmosphere and the ocean. How does a system respond to noisy forcing? The answer depends on its sensitivity.

STELLA offers a number of options for adding noisy forcing to a system. For the leaky bucket, noise can be added by including the factor

$$(1. + RANDOM(-1,1))$$

to the flow, Inflow. At each timestep, Inflow randomly takes on a value ranging from zero to twice its original value. The resulting Inflow is very noisy (Figure 1.9), but the variations in Depth driven by this noisy forcing are far smoother, as is the rate of outflow, which is determined by Depth. The water level represents a weighted average of Inflow over some past

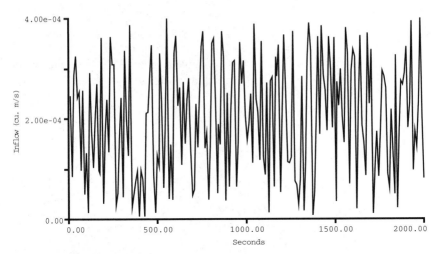

FIGURE 1.9   Noisy inflow to the leaky bucket.

time, and so a record of Depth gives a smoothed depiction of the history of Inflow. We saw this in Figure 1.6, in which a sudden turning on of the tap did not cause a sudden jump in Depth but, rather, a smooth rise. Similarly, the temperature of the earth responds smoothly to sudden variations in radiative heating, such as the decrease caused by a large volcanic eruption.

How does the sensitivity of a system affect its response to noise? We can again use "Sensi Specs" to run the model twice with a noisy inflow, once with a smaller hole (greater sensitivity) and once with a larger hole (less sensitivity). The results, shown in Figure 1.10, differ dramatically in their

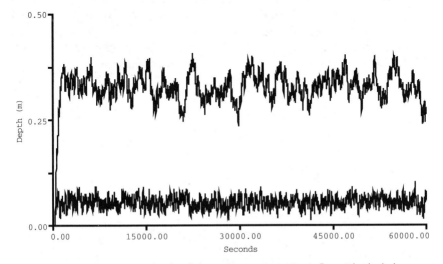

FIGURE 1.10   Variations in the depth in response to a noisy inflow. The hole has a radius of 5 mm for the upper curve and 8 mm for the lower curve.

appearance. Here, the top curve is the result with a 5-mm hole while the lower curve is that for an 8-mm hole. Not only does the more sensitive case show larger up and down swings in its response to the noise, but these swings last longer than for the less-sensitive model. The timescale of the variability, as well as its strength, depends on the sensitivity. This, too, has important implications for the climate, to which we return in Chapter 7.

Problems

1.1 It is asserted in Section 1.1 that the ball in a well conserves its energy. Verify this by adding converters to the model that calculate the kinetic energy of the ball, $1/2$ Velocity^2, and its potential energy $-1/2$ Constant $\times$ Position^2. The total energy is the sum of the kinetic and potential energy.

1.2 Explore how the radius of the bucket (as opposed to that of the hole) affects the sensitivity of the leaky bucket to inflow. Does the sensitivity of Depth to a change in Inflow depend on Radius_bucket? How are the time it takes for the model to equilibrate and the appearance of its response to noisy forcing affected?

Further Reading

There is a growing library of books on systems thinking and modeling, though many are written primarily for the business community. Two exceptions are *Dynamic Modeling,* by Bruce Hannon and Matthias Ruth (1994, Springer, 264 pp.), an excellent introduction to modeling dynamical systems using STELLA, and *Modeling the Environment: An Introduction to System Dynamics Modeling of Environmental Systems,* by Andrew Ford (1999, Island Press, 480 pp.), an introduction to systems thinking and modeling in the context of environmental issues.

2

Models of the Global Climate

2.1 No-Atmosphere Climate Model

In Chapter 1, it was asserted that the climate is a stable system, and we saw that stable systems are associated with negative feedbacks. Where, then, is the negative feedback in the climate system? It is contained within the simplest possible climate model, that for a planet with no atmosphere. Constructing this model introduces the units and dimensions used throughout this chapter.

In this simplest case, the temperature defines the climate. The temperature is a measure of the internal energy of the system, and it is, in fact, the energy that we want to keep track of in our model. While it is possible to model the energy of the entire planet, the units and values are more convenient if the planet is represented by an average square meter of its surface. Flows of energy are measured in units of watts (joules per second). In our model, we deal with flows of energy into and out of a square meter of planetary surface, so the units for flows are watts per square meter. The energy contained in the square meter depends on the temperature and the heat capacity, i.e.,

$$\text{Temperature} = \text{Energy/Heat capacity.}$$

The heat capacity of the system depends on the volume of the system, or, in other words, how deep a layer we wish to consider, and what it is made of. The ocean covers most of the earth, and in the ocean, heat is typically mixed rapidly to some depth. Therefore, the heat capacity is taken as that of a layer of water of this depth and is given by

$$\text{Heat\_capacity} = \text{Depth\_of\_mixed\_layer} \times \text{Specific\_heat\_of\_water}$$
$$\times \text{Density\_of\_water.}$$

The specific heat of water is 4218 (J/kg K), and its density is 1000 kg/m³.

Energy flows to the earth from the sun, primarily in the form of visible radiation. The flux of solar energy through a square meter oriented perpendicular to the sun's rays at the earth's average orbital distance from the sun is called the *solar "constant,"* even though in fact it varies slightly. An average value is 1367 W/m². The earth is not flat, however, so at most locations and times, the sun's rays strike its surface obliquely. Also, at any time, half the earth experiences nighttime and receives no solar energy. These two effects reduce the amount of solar energy impinging on an average square meter of the earth's surface by a factor of 4 from the solar constant. This may be understood by recognizing that the amount of solar energy intercepted by the earth is proportional to its cross-sectional area, which is one-fourth of its surface area. Furthermore, because the daytime side of the earth is visible from space, we know that the earth must reflect some of this solar energy. This fraction, called the *albedo,* is about 0.3. Taking these effects together, the solar energy absorbed by an average square meter of the earth's surface is given by

$$\text{Solar} = \text{Solar\_constant}/4 \times (1 - \text{Albedo}).$$

How does energy leave the earth? To travel into the vacuum of outer space, it must do so as electromagnetic radiation. All objects with temperatures above absolute zero radiate energy. The lower the temperature, the longer is the wavelength of this radiation. At typical terrestrial temperatures, the radiation is emitted in the infrared band, at wavelengths roughly 20 times that of visible light. Thus, the incoming sunlight and the outgoing infrared radiation may be treated as two completely distinct flows of radiant energy: one visible and ultraviolet; the other infrared. According to the Stefan–Boltzmann law, the energy radiated from a warm object—denoted IR, for infrared—is given by

$$\text{IR} = \text{Sigma} \times \text{T}^4,$$

where the temperature, T, is given in kelvins and Sigma is the Stefan–Boltzmann constant. This is a fundamental physical constant with a value of $5.6696 \ 10^{-8}$ W/(m² K⁴).

In the present context, the most important feature of the Stefan–Boltzmann law is that the warmer an object is, the faster it radiates away energy. For the earth, this provides the negative feedback needed to make the climate stable. In contrast, the rate at which the sun provides energy to the climate system does not depend on anything that happens on the earth (with the important qualification that the albedo may depend on the temperature, see Section 2.3). Thus, a negative feedback is operating on the outgoing infrared radiation and no feedback at all on the incoming solar radiation. The analogy with the leaky bucket (Section 1.2) is obvious.

Here is another extremely simple analog to the climate system that can be constructed in a few minutes: Take a cardboard box, place an electric lamp inside, and close the box. Insert a thermometer through a hole in the side of the box, so that it can be read from the outside (electronic thermometers work especially well for this). When the lamp is plugged in, the box warms up. The amount of energy coming into the box depends only on the wattage of the lightbulb and does not change with the temperature

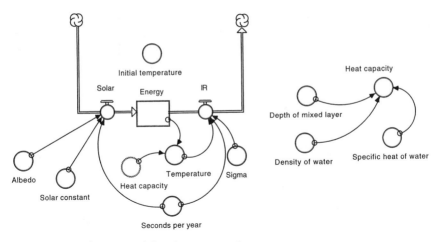

FIGURE 2.1   A climate model with no atmosphere.

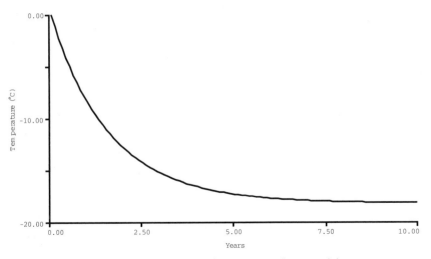

FIGURE 2.2   Temperature versus time in the no-atmosphere model.

of the box. But as the box warms, it loses heat more rapidly (by conduction, convection, and radiation) to the surrounding room. This is the crucial negative feedback. So, like the global climate, and like the leaky bucket, the box comes to a stable equilibrium temperature.

Returning to the climate system, our STELLA model is shown in Figure 2.1. When this model is run using 50 m for Depth_of_mixed_layer, it comes to a stable equilibrium in about 10 years (Figure 2.2). The stability of this equilibrium can be checked, as for the leaky bucket, by perturbing the temperature away from equilibrium value and watching it return. The equilibrium is indeed stable, but there is a problem with the final temperature itself. It is more than 18°C below zero! If the earth were this cold, life would be difficult if not impossible. What is wrong with the model? It has no atmosphere, and, as is demonstrated with the next model, it is the insulating effect of the atmosphere, denoted the "greenhouse effect," that makes the earth habitably warm.

2.2 One-Layer Atmosphere Model

How does the atmosphere insulate the surface of the earth? An insulator reduces the rate at which heat leaves a body. If you throw a blanket over the box with a lightbulb, the temperature rises to a new, warmer equilibrium. The flow of energy into the box is unchanged by the blanket, so in equilibrium, the flow of energy out of the box must also be unchanged. But with a blanket covering it, the box must be warmer in order to lose energy at the same rate as before. Putting a blanket over the box is closely analogous to reducing the size of the hole in the leaky bucket.

Because energy flows from the earth in the form of infrared radiation, if the atmosphere is to act as an insulator, it must interfere with the outward flow of infrared radiation from the earth's surface. This is indeed what happens. Some of the gases in the atmosphere, primarily water vapor, carbon dioxide, methane, nitrous oxide, and ozone, absorb infrared radiation. According to Kirchoff's law, the ability of a substance to absorb radiation of a given wavelength is equal to its ability to emit radiation at the same wavelength. The ability to emit is measured relative to the emission of a perfect radiator, a so-called blackbody. (In the previous model, it was assumed that the surface of the earth is a perfect radiator, or a blackbody.) Thus, if the atmosphere absorbs a fraction, called the *absorptivity,* of the infrared radiation from the earth's surface, it will emit infrared radiation, both downward to the surface and upward into space, at a rate given by

$$\text{IR\_atmosphere} = \text{Emissivity} \times \text{Sigma} \times \text{T\_atmosphere}^4,$$

where, according to Kirchoff's law, Emissivity is equal to the absorptivity.

The new model (Figure 2.3) includes an atmosphere that absorbs some of the outward-flowing infrared radiation from the surface. This energy warms the atmosphere, which then radiates infrared radiation up to space and down to the surface. The surface of the earth is a blackbody, and it absorbs all of the infrared radiation coming from the atmosphere. To consider the effects of the atmosphere completely, we include some absorption of solar radiation by the atmosphere. The atmosphere does not, however, emit visible radiation, because it is much too cold. To emit visible light strongly requires a temperature near that of the sun, around 6000 K.

This model has a second stock for the energy of the model atmosphere. The temperature of the atmosphere is obtained, as for the surface, by dividing the energy by the heat capacity. The heat capacity of the atmosphere is, however, much less than that of 50 m of water. This means the temperature of the atmosphere can change much more rapidly than that of the sur-

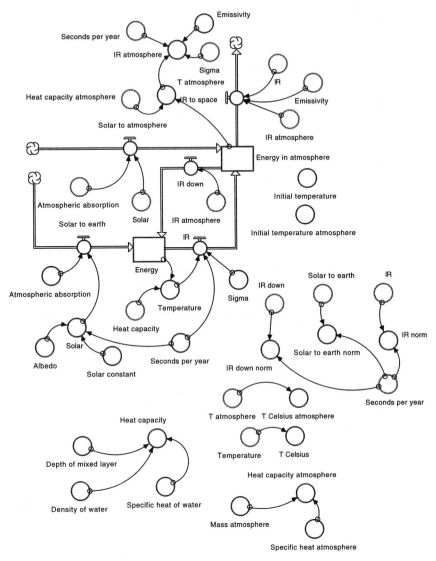

FIGURE 2.3   A climate model with a one-layer atmosphere. (IR, infrared radiation.)

face, and a much shorter timestep must be used than in the previous model. Otherwise the model will "blow up." The atmosphere has two flows of energy coming in: absorbed solar radiation (Solar_to_atmosphere) and infrared radiation from the surface (IR). Energy flows out of the atmosphere, both downward to the surface, IR_down, and out into space, IR_to_space. The latter flow includes infrared radiation from the surface that passes through the atmosphere without being absorbed, as well as that emitted upward by the atmosphere.

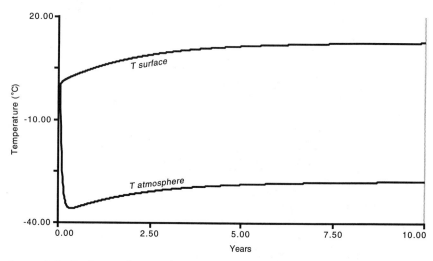

FIGURE 2.4   Surface and atmospheric temperatures in the one-layer climate model.

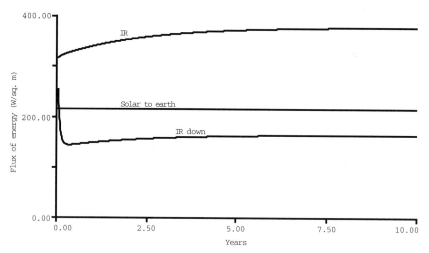

FIGURE 2.5   Flows of energy to and from the surface in the one-layer model.

If reasonable values are used for the infrared emissivity of the atmosphere and its absorption of solar radiation, 0.8 and 0.1, respectively, a reasonable and comfortable temperature, above 15°C, is obtained for the surface of the earth (Figure 2.4). Figure 2.5 shows the flows of energy arriving at and leaving the surface of the earth. The downward flux of infrared radiation from the atmosphere is nearly as important in warming the surface as solar radiation. Surprisingly, energy leaves the earth's surface as infrared radiation at a greater rate than it arrives from the sun at the top of the atmosphere. A unit of energy leaving the surface has only a 20% chance of reaching outer space directly. If the atmosphere absorbs it, it has a 50%

chance of being sent back to the surface. Thus, a unit of energy has a 40% chance of making at least one round-trip between the earth and the atmosphere before escaping to space. The upward flux of infrared radiation from the surface therefore includes energy that is effectively bouncing back and forth between the earth and the atmosphere.

What happens if the atmosphere absorbs as large a fraction of solar radiation as it does infrared? Then the atmosphere and the surface once again have the same unpleasantly cold temperatures as in the previous model. The atmosphere is an effective insulator for the surface of the earth only if it acts differently on visible and infrared radiation: it must be nearly transparent for visible radiation and nearly opaque for infrared. This is the origin of the term *greenhouse effect*. It was believed that the glass walls and roof of a greenhouse maintained an elevated temperature within by allowing solar radiation to pass through while absorbing the outgoing infrared radiation emitted by the warm interior. Now it is understood that the physical confinement of the air within the greenhouse—the warm air cannot rise and blow away with the wind—is actually more important than radiative effects, so the "greenhouse" terminology for the earth is something of a misnomer. The earth's climate is warmed by the "greenhouse effect" but the inside of a greenhouse is not.

As for the leaky bucket, we can see how sensitive the climate of this model is to changes in parameters. It is readily verified that this model is more sensitive to changes in the solar constant than the model with no atmosphere. This is analogous to how reducing the size of the hole makes the water level in the leaky bucket more sensitive to changing the inflow from the tap. How would the sensitivity of the leaky bucket change, however, if the size of the hole or the flow from the tap were made to depend on water level? Additional feedback mechanisms of this sort operate in the real climate system, and they are included in our next model.

2.3 One-Layer Atmosphere Model with Feedbacks

Much current interest in the climate stems from concerns about the anthropogenic greenhouse effect. Human consumption of fossil fuels—coal, oil, and natural gas—together with deforestation have caused atmospheric concentrations of carbon dioxide to increase by nearly 30% since the beginning of the Industrial Revolution, and half of that increase has come since 1960. It is almost certain that sometime in this century atmospheric levels of carbon dioxide will exceed 600 parts per million, more than double the preindustrial value. Doubling the atmospheric concentration of carbon dioxide has the direct effect, neglecting any other changes in the climate, of increasing the downward flux of infrared radiation at the ground by about 4 W/m^2. A similar increase in the solar radiation absorbed by the earth's surface would result from a 2% increase in the solar constant. In our one-level atmosphere model, such an increase in the solar constant raises the global temperature by less than 1.5 K. This is a modest change and probably would not be sufficient to engender much concern about environmental disruption from global warming. The problem comes with the phrase "neglecting any other changes in the climate." In fact, when the global temperature changes, other aspects of the climate change as well. These changes act as both positive and negative feedbacks for changes in the climate. The existence and complexity of these feedbacks make it difficult to project how the climate will change as humans add carbon dioxide and other greenhouse gases to the atmosphere.

Two of the most important changes can be represented, if crudely, in our one-layer atmosphere model. The first of these is the increase in atmospheric concentrations of water vapor as the temperature rises. Because the saturation vapor pressure of water depends strongly on the temperature, it is expected that there will be more water vapor in the atmosphere at higher temperatures. Water vapor is the most important absorber and emitter of infrared radiation in the atmosphere—the most important greenhouse gas—so, as the concentration of water vapor increases, the infrared emissivity increases as well. Therefore, the atmospheric greenhouse effect is stronger at higher temperatures. This is a positive feedback for climate change. It is akin to adding some mechanism to the leaky bucket that reduces the size of the hole as the water level rises.

The net emissivity of the atmosphere varies roughly with the logarithm of the water vapor concentration, while the water vapor concentration, at a fixed relative humidity, varies roughly exponentially with the temperature. Taken together, we end up with the following very approximate expression for the temperature dependence of the emissivity,

$$\text{Emissivity} = \text{MAX}(E\_0 + .1 \times T\_1 \times (1/T\_0 - 1/\text{Temperature}),0),$$

where T_0 is 273.15 K and E_0 is the emissivity when the temperature equals T_0. T_1 is a constant with the dimensions of temperature derived from the physical properties of water. The MAX, or maximum, function, is used to prevent the emissivity from becoming negative at low temperatures, which would be unphysical.

Another aspect of the climate that changes with the temperature is the fraction of the earth's surface that is covered by snow and ice. Because snow and ice are highly reflective, the albedo is higher when they cover more of the surface. As the temperature rises, the snow and ice line retreats poleward, decreasing the albedo, and thereby increasing the fraction of solar radiation absorbed by the earth's surface. Once again, then, this is a positive feedback for climate change. In the leaky bucket, this is like adding some mechanism that opens the tap as the water level rises.

A plausible dependence of the albedo on the temperature is shown in Figure 2.6. There is a maximum value, of 0.62, when the earth is entirely covered with ice and snow, and a minimum value, of 0.3, when the earth is entirely ice-free. The earth is assumed to remain ice covered below a criti-

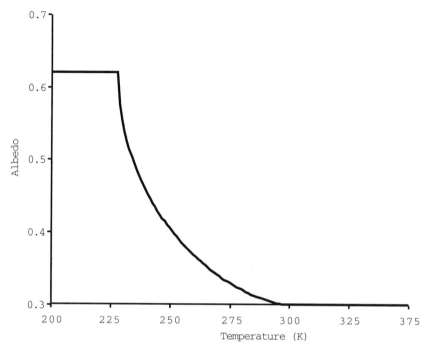

FIGURE 2.6   Temperature dependence of the albedo in a climate model with feedbacks.

cal global temperature, −45°C, so the curve is flat below that temperature. The curve is steepest at temperatures just above this critical value to reflect the fact that a given advance or retreat of the snow/ice line has the greatest effect on the absorbed solar radiation when this line is in the tropics, where the incoming solar radiation is strongest.

How do these feedbacks modify the sensitivity of the climate in this model? The model is run with each feedback turned on in turn. This is done by turning on the "switches," H2O_feedback and Albedo_feedback, setting the values in the converter to 1. The resulting changes in temperature in response to a 2% increase in the solar constant are (all the following values are in kelvins)

No feedbacks	1.43 K	—
Water-vapor feedback	2.44	1.01
Albedo feedback	1.62	0.19
Both feedbacks	3.01	1.58

When the water-vapor feedback or both feedbacks are turned on, the sensitivity of this model's climate—generally defined as the magnitude of the equilibrium warming obtained from doubling atmospheric carbon dioxide—is in the range of values obtained with global climate models. The second column gives the differences of the warming in each case from that obtained without feedbacks. The effect of combining the two feedbacks is greater than additive. Intuitively, this makes sense, in that the operation of each positive feedback mechanism amplifies the effect of the other mechanism. It is possible to predict quantitatively how multiple feedbacks will combine.[1] If the gain, g, is defined as the fraction of the total temperature change that is due to the feedback (e.g., the gain for water-vapor feedback is $g = 1.01/2.44 = 0.41$), then unlike the temperature changes themselves, the gains from different feedbacks are additive. The resulting temperature change is given by

$$\Delta T = \Delta T_{no\ feedback} /(1 - g),$$

where the gain, g, is the sum of all the gains from all the feedback mechanisms that are operating. The interesting and perhaps alarming conclusion from this simple expression is that several positive feedbacks combined can give rise to a total gain that is equal to or greater than 1, in which case the climate is unstable. The apparent stability of the earth's climate indicates that g is significantly less than 1, but perhaps this was not the case for Venus, whose climate succumbed to a "runaway" greenhouse effect.

The idea of gain is useful for dealing with the sensitivity or stability of the climate for small perturbations. There is the larger question, however, of whether the climate possesses just one or more than one stable states. Figure 2.7 shows the surface temperature for two sensitivity runs in which the starting temperature is varied from 200 to 300 K in intervals of 10 K. In both cases, albedo feedback is on, but in Figure 2.7a, water-vapor feedback is turned off, and in Figure 2.7b, it is on. In the absence of water-vapor feedback, all of the runs rapidly approach the same equilibrium temperature. When water-vapor feedback is on, however, the first three runs equilibrate at a much colder temperature, around 220 K ($-53°C$). At this temperature, the albedo is at its maximum value. In other words, this cold equilibrium state corresponds to an ice-covered earth. The final state of the remainder of the runs is close to what was obtained without water-vapor feedback.

To understand the existence of these two different equilibrium states, it is helpful to consider the fluxes of energy into and out of the climate system. These are plotted against the temperature in Figure 2.8. The *IR out* curves are obtained by increasing the solar constant so that the equilibrium temperature is greater than 300 K, then starting the model with a temperature of 200 K. When the model is run, the temperature then rises with time over a wide range of values. IR_to_space and the temperature are saved in a table. The *Solar in* curve is given by

Solar in = Solar constant/4 × (1 − Albedo),

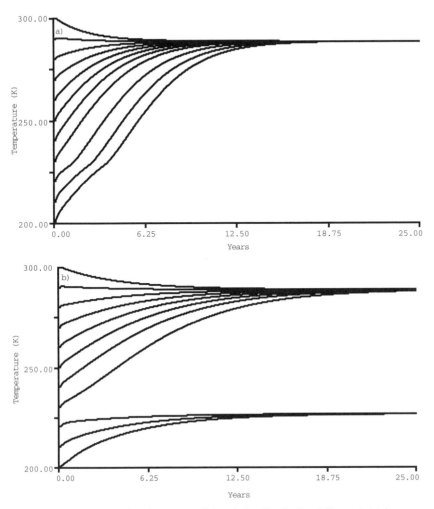

FIGURE 2.7   Results of the climate model with feedbacks for different initial temperatures. (a) Albedo feedback is on, and water-vapor feedback is off. (b) Both feedbacks are on.

where the albedo varies with temperature as before. In equilibrium, the energy fluxes into and out of the system must be in balance. In other words, equilibrium states occur where the *Solar in* and *IR out* curves intersect. Figure 2.8 shows that this happens at three temperatures when the water-vapor feedback is on but at only one temperature when it is turned off. This is consistent with the results of our sensitivity experiment, in that only one equilibrium is found when the water-vapor feedback is off. But why are only two equilibrium temperatures obtained with the water-vapor feedback on, whereas Figure 2.8 shows that energy balance can occur at

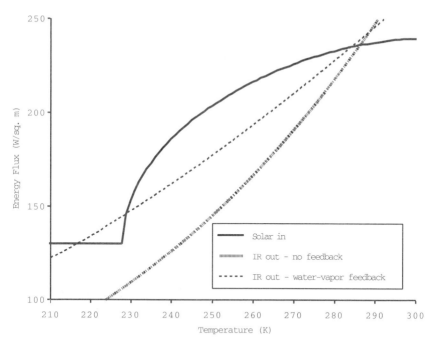

FIGURE 2.8   Fluxes of solar energy in and infrared radiation out as a function of the surface temperature in the model with feedback.

three different temperatures? The answer to this question lies in the stability of the equilibria. Consider the coldest intersection between the *IR out* and *Solar in* curves. When the temperature increases slightly, *IR out* increases more rapidly than *Solar in*. Thus, the net effect is a loss of energy that drives the system back toward equilibrium. The same is true for the warmest intersection, so these two equilibria are stable. For the middle intersection, however, *Solar in* rises more rapidly with temperature than does *IR out*. A positive perturbation of the temperature from this equilibrium leads, therefore, to net heating and further warming. In other words, the middle equilibrium is unstable, and a time integration of the model will not end in this state.

2.4 Radiative-Convective Model

Treating the entire atmosphere as a single layer is, of course, a gross idealization. Following the flows of visible and infrared radiation through the atmosphere with its inhomogeneous distributions of radiatively active gases, particles, and clouds is a hugely complicated problem and is the focus of the field of radiative transfer, on its own an entire scientific discipline. Within STELLA, it is, however, possible to make a first step toward consid-

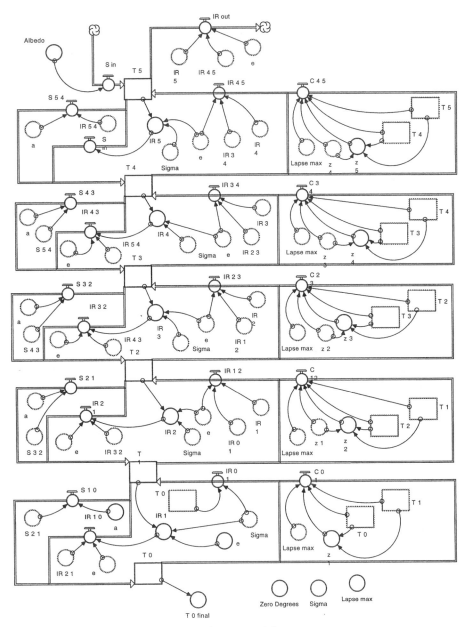

FIGURE 2.9   A radiative-convective climate model.

ering the flows of radiation through the atmosphere and by so doing to learn something about how the vertical profile of atmospheric temperature is determined.

Figure 2.9 shows a model in which the atmosphere has been broken into five layers; this number is readily increased and is limited only by the patience of the modeler. The atmosphere is treated as a stack of plain parallel slabs—only vertical transfers of radiation are considered. In a global climate model, many wavelengths of solar and infrared radiation would be treated separately, but here they are all lumped together. A temperature, an emissivity for infrared radiation, and an absorptivity for solar radiation characterize each layer. For this case, in the interest of simplicity, the solar absorptivity and infrared emissivity are assumed to be the same for each layer; this assumption is easily relaxed. Also for simplicity, the solar radiation that is lost to space by backscattering and reflection within the atmosphere and at the earth's surface is removed from the incoming solar radiation at the top of the model atmosphere.

With these assumptions, three streams of radiative energy flow through the model (the fourth stream of energy flow, represented by the C_- flows on the far right, is discussed below), down-going solar radiation, down-going infrared radiation, and outgoing infrared radiation. The solar radiation is the simplest. None is emitted within the atmosphere, and each atmospheric layer removes a fixed fraction. Some infrared radiation, both up- and down-going, is removed by each layer, but each layer also contributes to these flows through its infrared emissions. No infrared radiation is coming into the atmosphere from the top, and the upward stream begins with the emission from the earth's surface, which is assumed to be a blackbody. We are interested only in the equilibrium behavior of this model, not its time dependence, so for simplicity, the heat capacity of each layer, including the ground, is set equal to 1.

To represent the temperature profile as a function of altitude, it is necessary to know the altitude of each level. These are calculated using the principle of hydrostatic balance (see Chapter 3) and by assuming that the layers are equal in thickness in terms of atmospheric mass, or equivalently atmospheric pressure.

The resulting temperature profile is shown by the solid curve in Figure 2.10. A striking feature of this profile is that the temperature drops precipitously from the surface of the earth to the lowest atmospheric layer, which is centered at an altitude of less than 1 km. This is not realistic, and it suggests that something has been left out of the model.

The problem with this model is that it assumes that electromagnetic radiation accounts for all the vertical transport of energy within the atmosphere. In fact, energy is also transferred vertically by many different scales and types of atmospheric motions. Some of these motions, denoted *convection*, are driven by instabilities that occur when the temperature decreases too rapidly with increasing altitude. Models of this process are found in Chapter 3. For the present, it is sufficient to assume the existence of a "convective adjustment" that redistributes energy vertically in the atmospheric column whenever the temperature decrease with height exceeds a critical value. This redistribution is necessarily always an upward flux of energy and is accomplished by the C_- flows on the far right-hand side of the model in Figure 2.9.

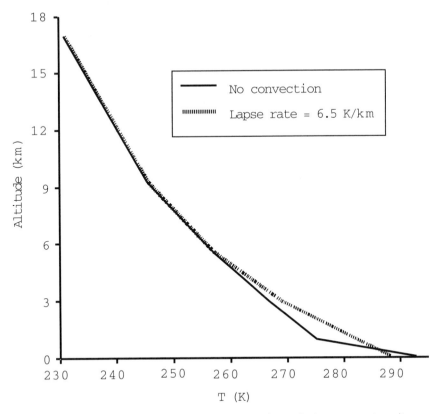

FIGURE 2.10   Vertical profiles of temperature in the radiative-convective climate model.

While the critical rate of temperature decrease with height, or lapse rate, that will cause instability depends on the humidity (see Chapter 3), for present purposes, it suffices to set this value as a fixed parameter in the model and to choose a value close to the observed lapse rate in the lower atmosphere, about 6.5 K/km. When the model is run including this additional flux of energy, the resulting temperature profile, the dashed curve in Figure 2.10, is much closer to what is observed.

This equilibrium solution permits examination of the energy balance at the surface of the earth. About 430 W/m^2 of energy arrives, roughly 60% of which is in the form of infrared radiation emitted by the atmosphere. The surface returns energy to the atmosphere in the form of heat and infrared radiation, in a ratio of about 1:10. This is an underestimate of the observed ratio, a consequence of the large layer thicknesses used in the model.

Because the maximum allowed decrease in temperature with height, Lapse_max, is a parameter in this model, it is worthwhile to see if the results

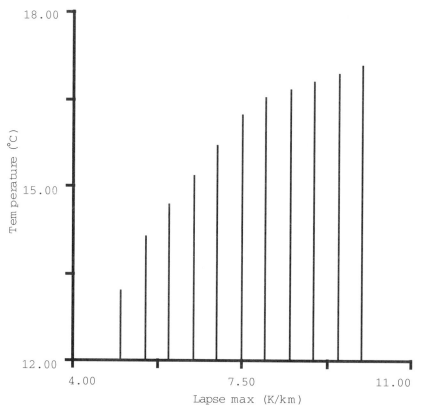

FIGURE 2.11   Sensitivity of the surface temperature in the radiative-convective model to changes in Lapse_max.

are sensitive to the value chosen. Figure 2.11 shows the results of a sensitivity run in which Lapse_max is increased from 5 to 10 K/km in increments of 0.5 K/km.

(A digression is needed here, to reveal the STELLA modeling trick required to produce a figure such as 2.11 from a sensitivity run. Define a new converter called, in this case, T0_final. This is connected to T0 and, with an IF-statement, is set to take on the value of T0 only at the final timestep of the simulation, and zero otherwise. T0_final is then plotted on a graph against the varied parameter, in this case Lapse_max, using the "Scatter" and "Comparative" settings in the graphing dialogue box.)

The surface temperature increases significantly with increasing values of Lapse_max. Why? For smaller values of the lapse rate, more energy must be transported by parameterized convection upward to levels where it is easily radiated directly to space. In other words, for a smaller lapse rate, energy is carried to levels of the atmosphere less well insulated from space by the blanketing greenhouse effect. This sensitivity to the lapse rate can lead to a

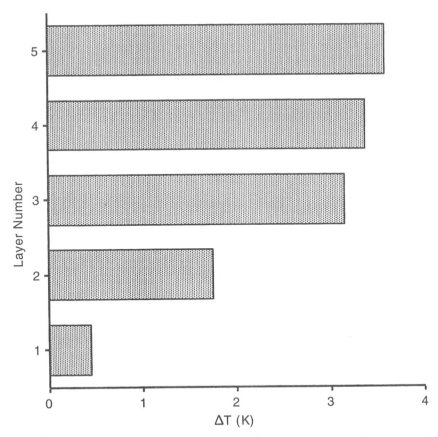

FIGURE 2.12   Changes in the surface temperature when the infrared emissivity of each level in the radiative-convective model is increased.

negative feedback for climatic change. As the climate warms, at least in global climate models, more vigorous tropical convection reduces the globally averaged lapse rate, and this increases the radiative cooling of the atmosphere to space.

With a model that resolves the atmosphere into several different layers, it is possible to investigate the influence of changing the infrared emissivity in individual layers. To do this, a small modification is made in the model to allow the layer emissivities, all equal to e in the original version, to take on different values. Increasing the emissivity, and hence the absorptivity, of a layer in essence increases the greenhouse effect due to that layer, so it is expected that the temperature will increase. Figure 2.12 shows the changes in surface temperature when each model layer in turn has its emissivity increased by 0.1, from 0.34 to 0.44. When this is done in layer 1, the temperature increases by less than 0.5 K, but when the identical increase in emissivity is applied to the top layer, the temperature increase is over 3.5 K.

Why? The higher layers in the model atmosphere are cooler, so when their emissivity is increased, the greater trapping of infrared radiation coming from the surface and lower atmosphere is relatively more important than their stronger emission of energy to space.

This is, again, a result that has important practical implications for climatic change. It suggests that the effect of an increase in the concentration of a greenhouse gas on the surface climate will depend critically on the vertical distribution of that gas. This is not an issue for carbon dioxide, which is relatively well mixed in the atmosphere. It is, however, an important issue for water vapor. As the climate warms due to the anthropogenic increases in the atmospheric concentration of carbon dioxide, the saturation vapor pressure for water also increases, so that, all other things being equal, there should be more water vapor in the atmosphere. Water vapor is the most important greenhouse gas, so it is expected that this effect is a positive feedback for climatic change. This is how water-vapor feedback functions in the single-layer model in the previous section. A complication is that most atmospheric water vapor resides within 1 or 2 km of the earth's surface, and, as has just been demonstrated, it is the water vapor in the upper troposphere that is most important in providing the expected positive feedback. If upper tropospheric water vapor were to decrease with increasing temperature, despite an overall increase in atmospheric water vapor, it would, in principle, be possible for the water-vapor feedback to be negative. Water-vapor feedback is positive in all current climate models, but the possibility that the existing models are wrong and that water-vapor feedback is negative has been raised by those who believe anthropogenic global warming will be less than the models predict. So far, however, no observational evidence indicates that the upper troposphere dries out as the climate warms, as would be required for a negative water-vapor feedback.

2.5 Advective Transport

Up to this point, we have discussed only the globally averaged climate without considering how energy moves horizontally within the climate system. Energy is transferred between the solid earth and the atmosphere as electromagnetic radiation, as heat, and, as we will see, as water vapor or latent heat. The fluids of the atmosphere and ocean carry energy from place to place. Before considering a model that includes the horizontal energy transports, it is worth considering two fundamentally different ways in which such transports can occur.

Advective transport occurs when heat, or some other property, is simply carried along on the wind or current. In a very simple depiction of this process (Figure 2.13), the wind blows through the model from left to right. If we imagine that the stocks are volumes of air at different points running west to east, then the model represents the transport of some property from

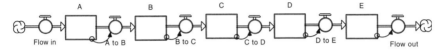

FIGURE 2.13   A model of advective transport.

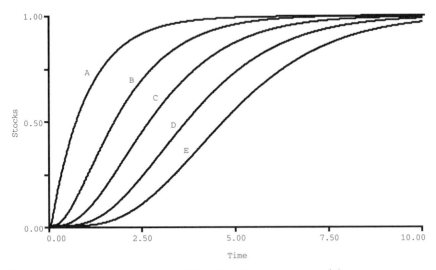

FIGURE 2.14   Levels in the stocks of the advective transport model.

west to east by the wind. Air blowing into the model from the west arrives
with a value of 1 for this property. As this air moves through the model, it
displaces air with a value of 0. The change in value from 0 to 1 moves
across the model approximately linearly with time. For simplicity, in this
model we have assumed that the wind has a constant value of 1, and that
each stock has a volume or capacity of 1. The smoothing of the 0-to-1 transition (Figure 2.14) as it moves through the model results from the large capacity of the stocks. The same thing happens in numerical models of
weather or climate because of the large size of the grid boxes, within which
the air is assumed to be well mixed. If we want the model to depict the ar-
rival of the sudden jump from 0 to 1 more accurately, we need only reduce
the volume of our stocks. Note that the wind carries the concentration of
the property, which is equal to the amount of the property in the stock divided by its volume—thus, the factor of 2 multiplying the flows in the new
model. Also, with the volume of each stock reduced by half, a concentration of 1 implies an amount in the stock equal to 1/2. The sharpness of the
transition is now better preserved, as may be seen by comparing Figures
2.14 and 2.15, but at the cost of needing twice as many stocks and flows as
before to model the same physical system. This simple example illustrates
the resolution dilemma for numerical models of weather or climate. To
preserve sharp changes, such as those that accompany a cold front, grid

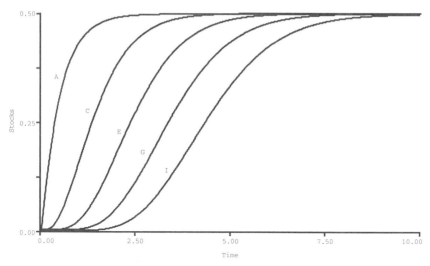

FIGURE 2.15   Levels in the stocks of an advective transport model with doubled resolution.

boxes must be small, but then the number of boxes needed to cover the earth is large, and the model requires either more computer time or a bigger computer.

2.6 Diffusive Transport

Advective transport is associated with the organized motion of a fluid from one place to another. *Diffusive* transport, by contrast, is associated with random motions either of molecules, as in the conduction of heat, or of bits of the macroscopic fluid, as in turbulent transport. Purely advective transport preserves the spatial patterns in the distribution of a property as it is carried downwind. Diffusive transport inevitably smooths such patterns. As was shown in the previous model, advective transport calculated at finite resolution invariably involves some inadvertent numerical diffusion. In STELLA, diffusive transport is represented by flows between stocks that are proportional to the difference in some property between the stocks. Figure 2.16 shows the diffusive counterpart to our simple advection model. The essential difference between the models is that the diffusive flows require infor-

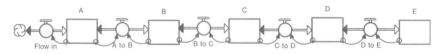

FIGURE 2.16   A model of diffusive transport.

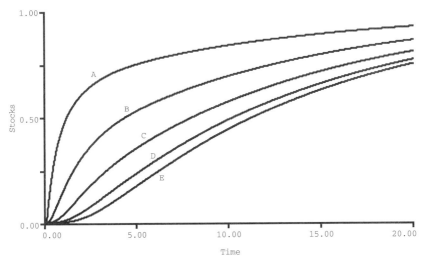

FIGURE 2.17   Levels in the stocks of the diffusive transport model.

mation about stocks on both sides, while advective flows depend only on the upstream stock. While the 0-to-1 transition was advected through the advective model at a steady pace, the diffusive advance of the same signal slows with time, and the fifth stock, E, approaches its final value only asymptotically (Figure 2.17).

2.7 Global Climate in a Shoebox

Using the idea of diffusive transport, we can build a conceptually simple model of the earth's climate, including the cycles of both energy and water. This model has the further advantage that a "hardware" model is readily constructed from common household materials. The reader is strongly encouraged to do so. To fix ideas, we describe a STELLA model with dimensions and parameters that describe our shoebox representation of the earth, rather than the actual planet. Figure 2.18 is a schematic of the shoebox model. A clear plastic shoebox has a lightbulb, the "sun," inserted through a hole in its side. Roughly half of the bottom of the shoebox is covered by a plastic bag of gravel, "the land," which slopes down to the other half, where there is a layer of water, the "ocean," a few centimeters deep. A small plastic cup, filled with ice water, is recessed in a hole in the lid of the shoebox. The portion of the cup that is inside the box is called the "cloud." When the sun is turned on, the interior of the box warms, and water evaporates from the ocean. Water vapor travels to the cloud, where it condenses and eventually drips down to the land as "rain." Heat is lost at the cloud by conduction through the sides of the cup. When this system is in equilibrium

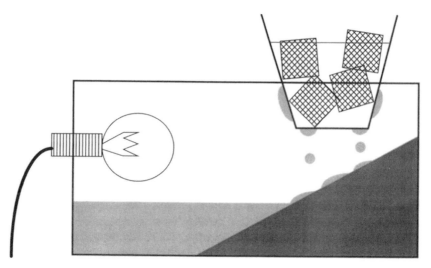

FIGURE 2.18   Schematic of a shoebox climate model.

(which requires that the ice in the cup be frequently replenished), the temperature inside the shoebox is steady, and there is a slow steady drip of water from the cloud. This water then runs off down the land back into the ocean.

How is this system like the earth? First, it is an open system for energy, which flows into the box, through the electric cord to the lightbulb, and out through the cloud, and also, but less importantly, through the sides of the box. Like the earth, the energy input to the shoebox is controlled externally, but the loss of energy depends on what happens inside. Like the earth, the shoebox is closed with respect to water, which endlessly cycles from the ocean to the air to the cloud to the land and back to the ocean. As in the earth's climate system, energy is transported within the shoebox as heat and as water vapor. These two forms of energy are denoted *sensible heat* and *latent heat*. Sensible heat is associated with the temperature of the air; it can be sensed. Latent heat is the energy that is stored in water vapor during evaporation. Latent heat is converted to sensible heat when and where the water vapor recondenses. On the earth, the poleward transport of energy as latent heat accounts for nearly half of the poleward transport of energy by the atmosphere across temperate latitudes.

Up to now, we have constructed STELLA models of highly idealized systems. Now we are confronted with a real physical system that we wish to represent in a STELLA model. When we look closely at our shoebox, we see that it is dauntingly complex. Rivulets of water run down the cloud, the air in contact with the sun is hotter than the air near the cloud, and, although we cannot see this, we can easily imagine that air is in motion everywhere within the box, rising where it is heated by the sun, and sinking where it is cooled by the cloud. The temperature and humidity at any one point in space and time are likely different from their values at every other point in space and time. To construct a manageably simple model of this system in STELLA, or, in fact, to construct a useful model of any real-world system using any computational or analytical tools, it is necessary, first, to understand which are the essential elements of the system and what basic principles must be represented in the model.

The STELLA model presented in Figure 2.19 is based on the following assumptions:

1. The box is sealed.
2. Energy enters the box only at the sun, and at a rate determined solely by the wattage of the lightbulb.
3. Water evaporates only from the ocean.
4. Heat leaves the box only at the cloud.
5. Water vapor condenses only at the cloud.
6. The temperature of the water in the ocean represents the temperature of the air in the box, and this is assumed to be uniform.
7. The temperature of the cloud is fixed at 0°C.

These are clearly idealizations, and this is by no means the only way a model of the shoebox could be constructed. The reader is encouraged to relax some of these assumptions, necessarily adding complexity to the model, and to see if such elaborations significantly change the modeled behavior.

The preceding points do not address the thorniest problem in constructing this model and, indeed, the most difficult aspect of modeling the earth's climate system. How do we represent the transports of heat and water vapor? In modern climate models, these transports are captured by modeling in detail the fluid dynamics of atmospheric motion. This approach is not feasible in STELLA (however, see Chapter 6), nor would it be especially helpful for understanding our shoebox. Rather, we make the simplest possible assumption, that the flows of heat and water vapor are diffusive. We assume that the air far away from the cloud has a uniform temperature, equal to that of the ocean, and that the temperature of the surface of the cloud is equal to that inside the cup, namely the freezing point of water. Heat flows down the gradient of temperature, at a rate proportional to the difference in temperature between the rest of the box and the cloud. The flow of sensible heat into the cloud is then given by

$$\text{Conduction} = \text{Density\_air} \times \text{C\_air} \times \text{Delta\_T\_cloud} \\ \times \text{Area\_cloud} \times \text{V\_diffuse},$$

where Delta_T_cloud is the difference in temperature between the rest of the box and the cloud, Area_cloud is the surface area of that portion of the cup that is within the box, and V_diffuse is a diffusion constant, expressed

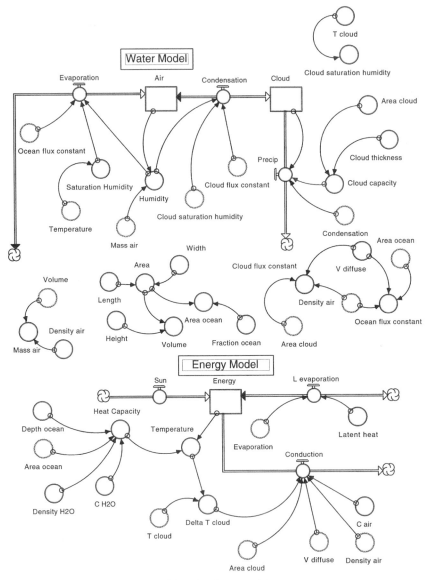

FIGURE 2.19   A representation of the shoebox model in STELLA.

in units of velocity. There is little to guide us in picking a value of V_diffuse, so it is essential that we check the sensitivity of the model to our choice.

Water vapor, or latent heat, is treated in much the same way as sensible heat. The air away from the cloud is assumed to be saturated (i.e., at 100% humidity), as is the air at the surface of the cloud (but here it is saturated at the temperature of the cloud). The specific humidity, the mass of water vapor per unit mass of air, at saturation is approximately an exponentially increasing function of the temperature. This is a consequence of the Clausius–Clapeyron equation, and our model employs an approximate solution of this equation. Because the specific humidity is greater elsewhere inside the box than at the surface of the cloud, water vapor diffuses toward the cloud, where it condenses.

Implicit in this discussion is the idea that our model of the shoebox comprises two different models, one for the flow of energy, and the other for the flow of water through the system. These two models are coupled in two ways. The energy model influences the water model, because the energy in the system determines the temperature, and the temperature determines, by way of the Clausius–Clapeyron equation, the humidity. The water model affects the energy model, because when water evaporates from the ocean, the box is cooled. When the water condenses on the cloud, the latent heat then released is taken up by the cloud and lost from the system. Thus, the L_evaporation flow in the energy model is proportional to the Evaporation flow in the water model.

A few more details in the model warrant explanation:

The water model is connected to STELLA clouds at both ends, even though we claimed earlier that this system is closed with respect to water. This is simply because we have assumed that at all times almost all the water in the system is in the ocean. Therefore, the volume of water in the ocean does not change significantly and may be treated as infinite. The ocean can easily be incorporated as an additional stock of water, if one wishes.

The energy is related to the temperature by the heat capacity of the ocean only, and the energy in the atmosphere is neglected. This approximation is valid as long as the heat capacity of the ocean is much greater than that of the air in the box.

It has been assumed that the cloud can only hold a certain volume of water. Once the water in the cloud exceeds this volume, all additional condensation is balanced by precipitation.

When the model is run with the parameters supplied, the temperature approaches its equilibrium value in between 1 and 2 hours. At this time, it is readily verified that the flows of energy out of the system by evaporation and by the conduction of sensible heat balance the energy input from the sun. These two energy outflows make roughly equal contributions toward balancing the solar input. The flows in the water model are similarly in balance. Indeed, the water budget achieves a nearly perfect balance within a few minutes, and the long equilibration time of the model is strictly that required to warm the ocean. In other words, because heat resides in the air and in the ocean (but mostly in the ocean), while water vapor resides only in the air, the water budget is at any time nearly in a state of equilibrium appropriate to the temperature at that time.

A number of instructive experiments can be performed with this model to determine which parameters control the temperature and why. Of particular interest is the sensitivity of the climate of the shoebox to changes in its solar constant (the wattage of the lightbulb). As the wattage is increased in equal increments, the equilibrium temperature also increases, but by a decreasing amount (Figure 2.20a). In other words, as the sun gets stronger, the climate in the shoebox gets warmer, but at a decreasing rate. The source of this decreasing sensitivity is the nonlinearity inherent in the Clausius–Clapeyron equation. As the climate warms, the humidity of the atmosphere increases nearly exponentially, the flow of water vapor to the cloud increases proportionally, and the latent heat (Figure 2.20b) becomes more and more important relative to the sensible heat in the energy budget (Figure 2.20c).

The shoebox leads to a correct prediction about the climate of the earth, that the role of latent heat in the transfer of energy is generally greater in a warmer climate or region or time of year than in a colder one. The model leads to a second prediction, however, that is completely wrong. This is the

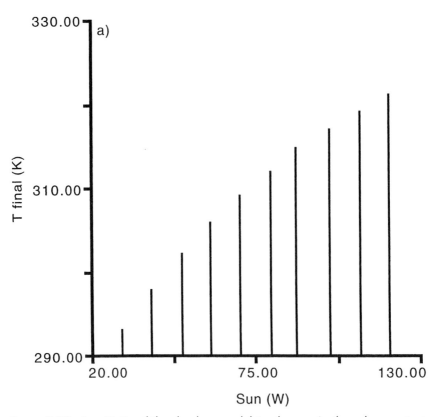

FIGURE 2.20   Sensitivity of the shoebox model to changes in the solar constant: (a) temperature; (*continued*)

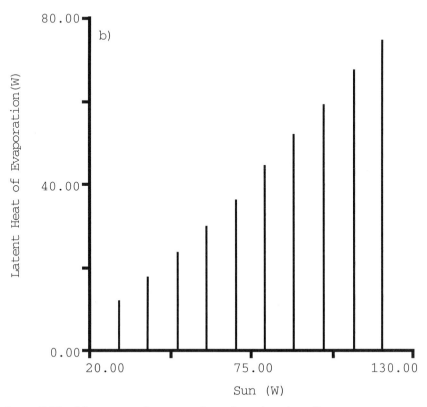

FIGURE 2.20   (b) transport of energy as latent heat; (*continued*)

prediction that, because the exponential increase in evaporation with temperature limits temperature increases, climate sensitivity to changes in external forcing decreases with increasing temperature. (Interestingly, a well-known climate scientist made precisely this error in the 1970s, calculating a very small sensitivity of the climate to anthropogenic increases in greenhouse gases.[2]) This wrong prediction results from a feature of the shoebox that is not shared by the real climate system, the presence of a perfect "heat sink" held at a constant temperature—the cloud. In the real climate, if the temperature warms, it is indeed true that there will be more water vapor in the atmosphere and that this will increase the efficiency with which energy is transported poleward. Unlike the cloud in the shoebox, temperatures at high latitudes are not held fixed. Rather, they change as is necessary for the radiative loss of energy to space to balance the poleward flow of energy

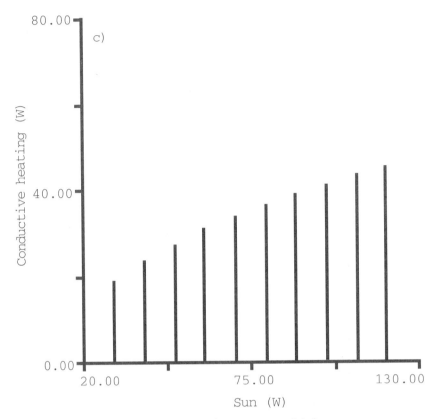

FIGURE 2.20   (*continued*) (c) transport of energy as sensible heat.

from lower latitudes. Climate models that capture this behavior in the simplest possible way are called *energy-balance climate models* (EBCMs). Our
next model is an example.

2.8 Energy-Balance Climate Model

The idea behind the EBCM is that the transfer of heat between different latitudes can be treated as a diffusive process. We consider only the meridional transfer of heat, because at the level of simplification of such a model,
the climate is assumed to be the same at all points at the same latitude. At
the very basic level, it is known that the processes that move heat and
moisture in the atmosphere tend to move heat from warm regions to cold
regions and move moisture from humid regions to dry ones, so in this
sense, the diffusive model is apt. There is, however, no reason that this

should be a linear diffusion, in which the transfer of heat is proportional to the difference in temperatures. Linear diffusion is assumed because it is the simplest possible representation of processes that in nature are extremely complex.

The present STELLA version of the EBCM (Figure 2.21) is restricted, in the interest of simplicity, to just two latitude bands. If these are chosen to be that between the equator and 30° and that between 30° and the pole, we have the convenience that their surface areas are the same. In this case, the average latitude of the area in the tropical region is nearly 15°, and that in the temperate box is about 51°. The transfers of heat and moisture (latent heat) across 30° are assumed to be proportional to the differences in average temperature and humidity between the regions north and equatorward of 30°.

We knew how much water vapor was in the air in the shoebox, since we assumed that it was saturated. This is not a good assumption for the atmosphere of the earth, which is saturated only in limited regions, such as inside clouds, and for short times. Instead, a value of 50% is chosen, rather arbitrarily, for the relative humidity (the ratio of the specific humidity to that at saturation). The transfer of moisture from warm to cold regions, as occurred in the shoebox, is preserved, however. It is assumed that an excess of evaporation over precipitation makes up the loss of vapor by transport from the tropics, and an excess of precipitation over evaporation balances the gain of vapor in temperate latitudes.

The EBCM could be constructed by adding some diffusive transfers of heat and moisture between two one-layer or multilayer atmosphere models. The disadvantage would be that the model would then require the short timestep needed for the atmosphere. If we are only interested in the surface climate, the role of the atmosphere in moderating the loss of energy to space by way of infrared radiation can be represented by a simple function of the surface temperature. This is called a *parameterization;* we wish to *parameterize* infrared radiation to space in terms of the surface temperature alone, thus avoiding any representation of the atmosphere in the model.

We can develop our parameterization using the one-layer atmosphere model described earlier in this chapter (Section 2.2). That model is run starting from a cold temperature, and we use the "Scatter" feature in STELLA's graphing dialogue window to make a graph of the outgoing infrared radiation as a function of the temperature. After a short initial period, when the temperature of the atmosphere adjusts to that of the surface, the flux of infrared radiation to space maintains a nearly linear relationship with the surface temperature. Furthermore, a sensitivity run reveals that the slope of this line depends on the emissivity of the atmosphere. If some of these values are saved in a table, the equation of the approximately linear relationship can be determined algebraically, and this linear relationship can be used to parameterize the radiative effects of the

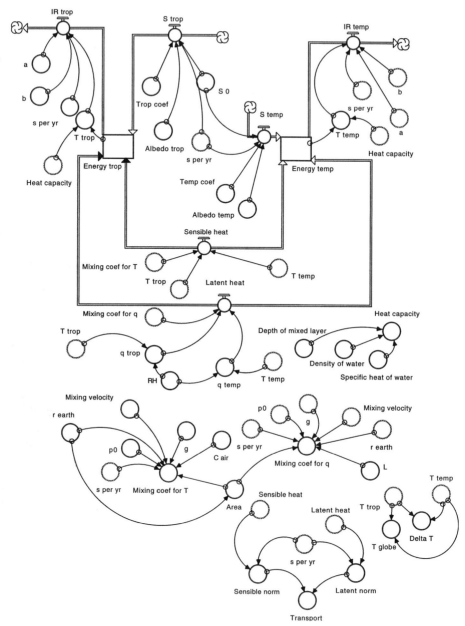

FIGURE 2.21   An energy-balance climate model.

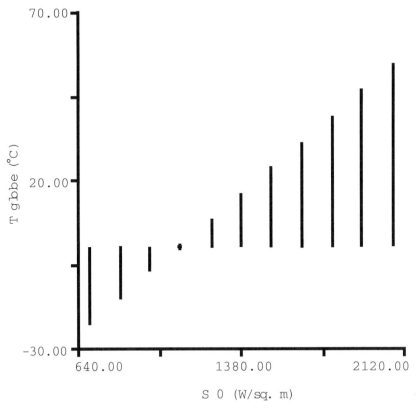

FIGURE 2.22   Sensivity of the energy-balance climate model to changes in the solar constant.

atmosphere in our EBCM. The resulting parameterized infrared cooling rate is given by

$$IR = a \times T + b,$$

where $a = 3.18$ W/[m^2 K], $b = -699.8$ W/m^2, and the temperature, T, is given in kelvin. The negative value of b is not meant to imply that infrared warming takes place at low temperatures. Rather, it is a consequence of using a straight line to represent a dependence that is nonlinear when considered over a wider range of temperatures than are likely to occur in our model.

When our EBCM using this parameterization is run with different values of the solar constant, the sensitivity of the globally averaged temperature to changes in solar constant is quite linear (Figure 2.22) over a wide range of solar constants, varying from 50% to 150% of the standard value. The apparent saturation of the temperature in the shoebox was an artifact of the "cloud" that acted as a perfect heat sink. An interesting feature of this model, however, is that the difference in temperature between the tropics and temperate latitudes decreases with solar constant when the climate is sufficiently warm (Figure 2.23). This results from the increased effectiveness of latent-heat transport at higher temperatures and is, once again, a consequence of the nonlinearity of the Clausius–Clapeyron equation. The association of warm climates with a weak equator-to-pole temperature contrast has been noted in the paleoclimatic record. For example, the earth is believed to have been much warmer during the Cretaceous Period than now, perhaps because of higher concentrations of carbon dioxide in the atmosphere, and it is also thought that the equator–pole temperature difference during the Cretaceous was much smaller than today.

Because the EBCM includes different geographical regions, albeit only two in this case, this model can be used to look at the seasonality of the cli-

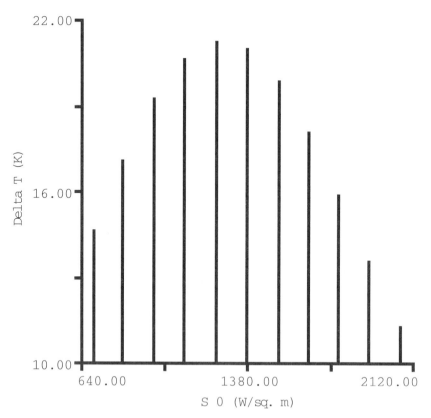

FIGURE 2.23   Sensitivity of the tropical–extratropical temperature contrast in the energy-balance climate model to changes in the solar constant.

mate. The results from such a seasonally varying EBCM make more sense if we first examine a much simpler system with periodic forcing.

2.9 Simple Periodic System

Consider the very simple model shown in Figure 2.24. A periodic flow, one that varies sinusoidally with time (this must be a biflow), feeds a stock (which must be allowed to take on negative values). The second flow out of the Stock, denoted Damping, is proportional to the value of the Stock. The change with time in the level of the Stock (denoted Storage) must at all times balance the sum of Periodic_input and Damping. If Damping_rate is very small (say 0.5, Figure 2.25a), then Storage balances Periodic_input. When the input is at its maximum, the level of the Stock is rising most rapidly, and when the Periodic input has returned to zero, the level of the Stock is at its maximum, and its rate of change, Storage, is equal to zero. The result is that the level of Stock is a sinusoid that lags behind Periodic input by a quarter of a cycle. If, on the other hand, Damping rate is large (say 50, Figure 2.25b), Damping largely balances Periodic_input. Since Damping is proportional to the level of the Stock, the level of Stock tracks Periodic_input with little lag. In this case, the damping timescale is short compared with the period of the oscillatory forcing, and the system essentially moves through a series of periodically varying steady states. In general, the lag between Periodic_input and the level of Stock will vary between zero and one quarter cycle. Smaller lags are associated with

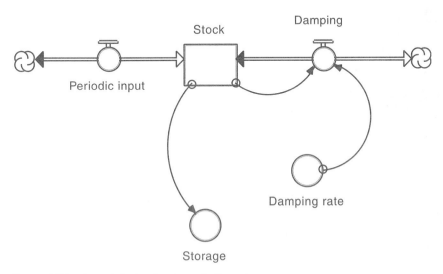

FIGURE 2.24   A model of a simple periodic system.

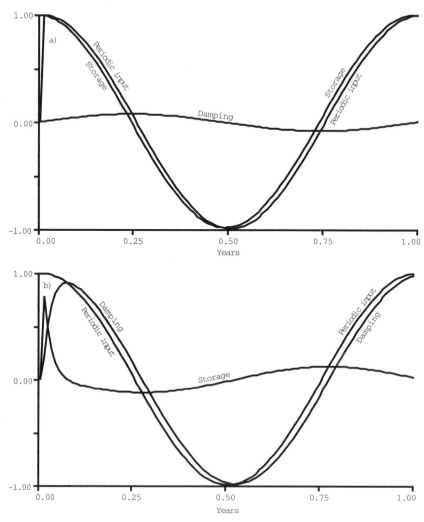

FIGURE 2.25   Flows in the simple periodic system with (a) weak damping and (b) strong damping.

higher Damping_rate and the reduced importance of Storage in the instantaneous budget of Stock. Figure 2.26 shows a single cycle of the level of Stock for runs with values of Damping_rate ranging from 0 to 10. As the damping increases, the amplitude of the fluctuations in Stock decrease, but they become more and more in phase with the fluctuations in Periodic_input. As we will see in the next model, the land and the atmosphere have only a small heat capacity, and at any time during the year, the average temperature represents a nearly instantaneous balance among the

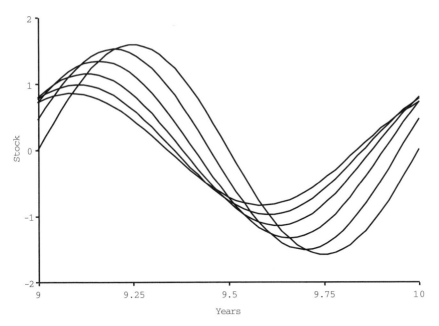

FIGURE 2.26   Levels of Stock in the periodic system for values of Damping_rate ranging from 0 to 10.

various fluxes of heat into and out of that region. A small heat capacity reduces the damping time and so is equivalent to the strong damping case in the present model. Over the continents, then, the temperature is nearly in phase with the periodic annual variations in solar forcing. In the ocean, however, the storage of energy from one season to the next is significant, and the temperature lags nearly a full season behind the seasonal variations in solar heating.

2.10 Energy-Balance Climate Model with Seasons

The seasonal cycle is very different, in both magnitude and phase, on land from that over the ocean. Figure 2.27 shows a model on an idealized earth on which the tropics are entirely ocean covered, and on which continents occupy half of the surface area of the extratropics. A number of assumptions are required. The heat capacity of the land is assumed to be just that of the atmosphere, which is equal to that of a layer of water about 2.5 m deep. The oceans are assumed to have the heat capacity of 50 m of water. The atmosphere transports heat from the tropics to the extratropics and between the land and ocean in the extratropics. For simplicity, the transports of latent and sensible heat are combined in this model. Because prevailing

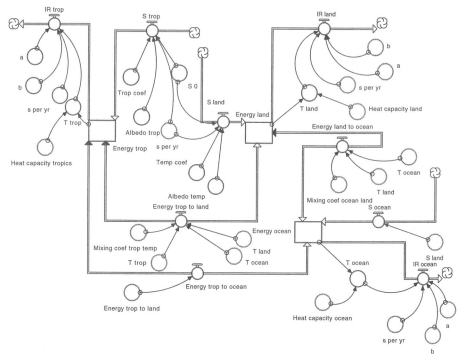

FIGURE 2.27   An energy-balance climate model for an idealized earth with a seasonal cycle.

winds in the extratropics are primarily west-to-east, it is assumed that the transport between the ocean and the land is twice as rapid as that from the tropics to the extratropics. Finally, the transport of heat from the tropics to the extratropics is assumed to be the same for the land and the ocean and to depend on the average temperature of the extratropics. This assumption is justified by the observation that storms (see Section 6.2) that develop over the continents travel over the ocean, where they continue to carry heat northward.

The seasonal cycle in this model is imposed by a periodic variation in the solar radiation. The magnitude of this variation is assumed to be much greater in the extratropics than in the tropics. A great many complexities could be added to this model that are left out. These include the difference in albedo between the land and the ocean and, because of the high albedo of snow and ice, the temperature dependence of the albedo on land.

Figure 2.28 shows 2 years of the annual march of temperatures generated by this model. The annual range of temperatures over the land is about twice as great as over the ocean. The land is coldest soon after the start of each year—set for convenience to be the winter solstice or the annual minimum in insolation in this model, while the ocean is coldest

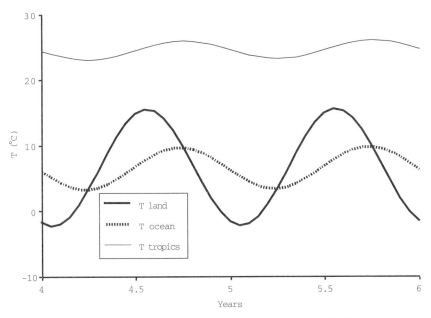

FIGURE 2.28   Annual cycles of temperature in the model of an idealized earth.

nearly a quarter of a year later. In relation to the previous simple periodic system, the land corresponds to a system in which the damping is much greater than the storage, while for the ocean the opposite is true. This is confirmed by considering the energy budgets of the land and the ocean. For the land (Figure 2.29a), the energy budget is nearly in instantaneous balance at all times. In winter, strong net radiative cooling balances the transport of heat to the land from both the ocean and the tropics, while in summer, there is weak net radiative heating, and energy arriving from the tropics is exported to the ocean. For the ocean (Figure 2.29b), the storage of heat is critically important; this is the term with the largest amplitude annual cycle. In the winter, the release of heat that has been stored during the summer, together with the transport of heat from the tropics, balances both the radiation loss of energy and the export of heat to the land. In summer the heat to be stored comes from radiation, the land, and the tropics. The wintertime release of heat from the ocean implies an upward transfer of energy across the air–ocean boundary, warming the atmosphere strongly from below. As will be seen in Chapter 3, this will tend to make the atmosphere unstable and, therefore, to promote storminess. Indeed, the extra-tropical oceans are stormy places in the winter, whereas in the summer, when the ocean is cooling the lower atmosphere, stable domes of high pressure dominate the oceanic climate.

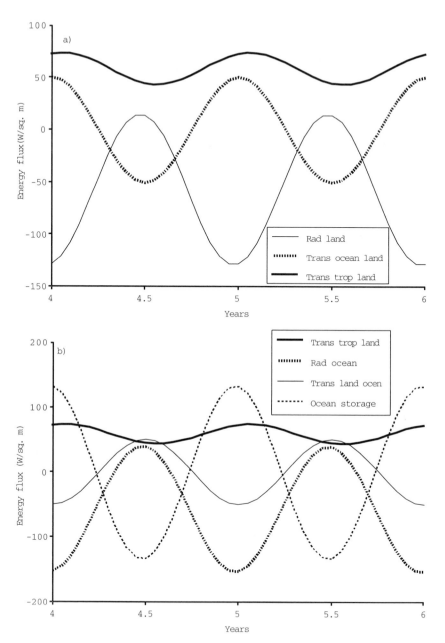

FIGURE 2.29   Energy budgets for the model of an idealized earth: (a) land, (b) extra-tropical ocean.

Problems

2.1 It is now generally believed that the impact of an asteroid or comet on the earth was responsible for the extinction of the dinosaurs and many other species 65 million years ago at the end of the Cretaceous Period. While there are several different ways that an asteroid impact could exterminate most life on the earth, one idea is that the cloud of dust and soot (from fires ignited by the impact) caused severe global cooling—a prolonged "asteroid winter." This idea also appeared in the 1980s ("nuclear winter") as a prediction for the likely climatic consequences of a nuclear war. To explore this idea, insert into the one-layer atmosphere model (Section 2.2) a thick cloud with a solar absorptivity close to 1, but keep the infrared absorptivity unchanged. Do your results substantiate the "asteroid winter" hypothesis?* What if the infrared absorptivity also increases?

2.2 Section 1.2 discusses how the time-dependent behavior of a system is related to its sensitivity. The one-layer climate model (Section 2.3) can be run with no feedbacks, with just water-vapor or snow/ice-albedo feedback or with both, the latter model being the most sensitive, and the former the least. Perturb the equilibrium states of each of these models by 1 K, and observe their return to equilibrium. Is the difference in approach to equilibrium what is expected?

In Section 1.2, it was mentioned that the response of a system to noisy forcing depends on its sensitivity. Use the RANDOM function of STELLA to add a small amount of noise to the solar forcing in the one-level model and observe the time-dependent response. Repeat this experiment with the feedbacks turned on and off. Do the runs differ in the way you expect?

The water-vapor feedback can be strengthened by using a value larger than 0.1 in the Emissivity converter. Confirm that this model does indeed become unstable when the total gain is greater than 1. How is the time-dependent behavior of the model altered when the total gain is less than but nearly equal to 1?

2.3 Try inserting a cloud into different layers of the radiative-convective model (Section 2.4). Clouds increase the albedo and, to a good approximation, do not absorb any solar radiation. Their emissivity in the infrared is close to unity (but remember that the global cloud cover is always much less than 100%). Can you confirm the observation that high clouds tend to warm but low clouds tend to cool the climate?

---

*This question has been posed on examinations by Professor Michael Schlesinger, my colleague in the Department of Atmospheric Sciences at the University of Illinois.

2.4 The models on the CD-ROM accompanying this book have been set to run with timesteps, denoted DT in STELLA, that are sufficiently short to produce reasonable results. When one constructs models on one's own or modifies existing models, it is easy to obtain surprising and apparently nonsensical results, for example, from using a timestep that is too long. This is called numerical instability. Try running the advective transport model (Section 2.5) with increasing values of DT. How does the character of the solution change, especially when DT is made equal to or larger than 1? If the timestep is further reduced from values much less than 1, does the character of the solution change? Is the critical value of the timestep at which the solution becomes utterly nonsensical different for the higher-resolution version of the model?

2.5 Explore the sensitivity of the equilibrium state of the shoebox model (Section 2.8) to parameters other than the strength of the lightbulb, namely, the area of the "cloud," Area_cloud, and the strength of diffusive transport, V_diffuse. In each case, predict in advance whether increasing or decreasing these parameters will make the climate warmer or cooler and whether it will increase or decrease the strength of the model's hydrologic cycle. Then perform your experiment, and test your prediction. If the results contradict your prediction, try to locate where your reasoning went astray. This will generally require checking the sensitivity of model outputs other than the temperature and precipitation rate.

2.6 How would the seasonal energy balances and temperatures of the extratropical land and ocean change if years on earth were either much longer or much shorter than they are? This can be changed in the seasonal EBCM (Section 2.10) by increasing or decreasing the value of s_per_yr. As before, predict the results before you do the experiment, and see if your predictions hold true.

Further Reading

The energy balance of the earth is discussed in *Global Physical Climatology,* by Dennis L. Hartmann (1994, Academic Press, 411 pp.). Details of the observed budgets of energy and water in the climate system are provided in *Physics of Climate,* by José P. Peixoto and Abraham H. Oort (1992, American Institute of Physics, 520 pp.).

References

1. Hansen, J., A. Lacis, D. Rind, G. Russell, P. Stone, I. Fung, R. Ruedy, and J. Lerner, 1984: Climate sensitivity: analysis of feedback mechanisms, in

Climate Processes and Climate Sensitivity, J. E. Hansen and T. Takahashi, eds., American Geophysical Union, Washington, D.C., 368 pp.

2. Newell, R. E., and T. G. Dopplick, 1979: Questions concerning the possible influence of anthropogenic CO_2 on atmospheric temperature. *Journal of Applied Meteorology,* **18**, 822–825.

3

Thermodynamics and Dynamics in the Vertical

3.1 Hydrostatic Balance

Much can be learned about the local weather—the height at which clouds will form, the likelihood of thunderstorms or of an air-pollution episode—by analyzing the thermodynamics and dynamics of parcels of air moving up and down through the atmosphere. Such parcels move through an atmospheric environment in which the vertical forces, the weight of the air and the upward decrease in pressure, are in balance. This is termed *hydrostatic balance*. The first model in this chapter uses this balance to find the pressure and density of the atmosphere given the pressure at the ground and a vertical temperature profile. This model then becomes a component of the subsequent models, which include first the thermodynamics, dry and moist, and then the dynamics, of air moving vertically.

When a fluid is in hydrostatic balance, the pressure at any height is equal to the weight, per unit area, of all the fluid in the overlying column. At increasing altitudes or decreasing depths, there is less mass overhead, and the pressure necessarily decreases. In a fluid with a constant density, such as, to a good approximation, water, the decrease in pressure with height is linear. In a gas, however, the density varies with the pressure. The atmosphere is very nearly an ideal gas, and in an ideal gas, the density is given by

$$\rho = p/(RT),$$

where p is the pressure in pascals, T is the temperature in kelvins, and R is the gas constant for the gas in question, 287 J/(kg K) for dry air. The decrease in density with decreasing pressure results in a positive feedback, by way of the density, between the pressure change with height and the pressure itself, as is shown in this model (Figure 3.1). The model is constructed so as to scan the profiles of temperature, pressure, and density upward with time. Here, the vertical velocity is imposed, though in later models in this chapter it will be determined by Newton's second law. As the altitude increases, the positive feedback between pressure and density causes the decrease in pressure to be faster than linear. The model assumes a linear

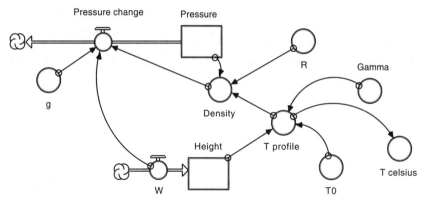

FIGURE 3.1   A model of hydrostatic balance.

decrease in temperature with the height, where Gamma denotes the rate of decrease with height, commonly called the *lapse rate*. In the case of a constant temperature (Gamma = 0), the pressure drops exponentially with increasing altitude. The density is lower, and the pressure therefore drops more slowly, when the temperature is higher. This is illustrated in Figure 3.2,

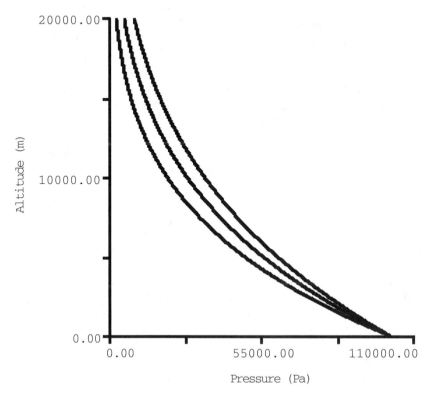

FIGURE 3.2   Atmospheric pressure profiles for three different values of the lapse rate, Gamma.

which shows profiles of pressure for three different values of Gamma, with the smallest value of Gamma corresponding to the rightmost curve. The surface temperature (T0) is the same for all three curves. The pressure drops most rapidly with height when Gamma is greatest and the temperatures aloft are lowest.

3.2 Dry Thermodynamics

Hydrostatic balance describes the environment through which parcels of air rise and sink. Before discussing the thermodynamics of a parcel, it is necessary to define what a parcel is. A *parcel* is a hypothetical blob of air sufficiently small that we can imagine it moving up and down through the atmosphere but large enough to have a temperature and humidity distinct from its environment. A parcel achieves a pressure equal to that of its surroundings in a time proportional to the speed of sound and inversely related to the size of the parcel. This is very fast compared with the amount of time required for other properties of the atmosphere to become uniformly distributed. One consequence of the rapid equilibration of pressure, compared with that of temperature or humidity is that a satisfactory measurement of atmospheric pressure can be obtained from a barometer indoors, while for meteorological purposes, a thermometer or hygrometer must, of course, be outside. Over the time spans of minutes that it takes for a parcel to rise or fall through the atmosphere, it is reasonable to assume that it maintains the same pressure as its environment, but that it does not exchange heat or water vapor with its environment. A later refinement of this idea is that any liquid water formed by condensation is assumed to immediately fall out of the parcel as rain. This adiabatic, or, more correctly, pseudo-adiabatic, assumption is the basis for understanding the behavior of rising or sinking air in the atmosphere. We begin by considering the thermodynamics of a dry parcel rising through a hydrostatic atmosphere.

As a parcel of air rises through the atmosphere, it is subject to falling pressures, to which it responds by expanding. In so doing, it does work on the surrounding atmosphere. The energy required to do this work is extracted from the internal energy of the parcel, given by $c_V T$, where c_V is the specific heat of dry air at constant volume, 717 J/(kg K). The work done by the parcel is equal to the pressure times the change in its volume. To our earlier model, we now add a stock to keep track of the internal energy with a single flow representing the work done by the parcel on its surroundings (Figure 3.3). For simplicity, a parcel of unit mass is assumed, so that the volume is equal to the specific volume, which is in turn equal to the reciprocal of the density. As the parcel rises, it cools at a rate, with respect to height, that in a hydrostatic atmosphere is nearly constant (Figure 3.4). It is readily derived analytically that this rate of cooling with height is given by g/c_p, where c_p, the specific heat at constant pressure, is given by

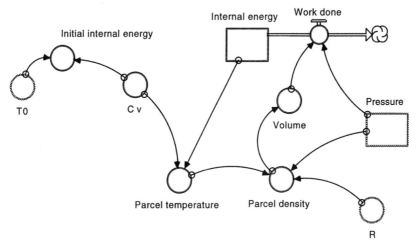

FIGURE 3.3   A model of the internal energy of a dry parcel of air.

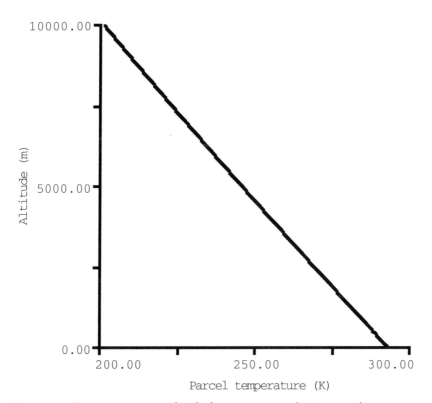

FIGURE 3.4   Temperature versus height for an unsaturated rising parcel.

$c_v + R = 1004$ J/(kg K) for dry air. This yields a value, denoted the dry-adiabatic lapse rate, of about 10 K of cooling for every kilometer a parcel rises. This is the rate at which a rising parcel cools, and it should not be confused with the rate, earlier denoted Gamma, at which the temperatures in the surrounding atmosphere decrease with increasing altitude. The parcel and the changes in its temperature with altitude are theoretical constructs. The environmental temperature profile, on the other hand, is what is observed whenever a weather balloon takes a sounding of the atmosphere.

It is readily verified in this model that, while the temperature of a rising or falling parcel of air changes, a quantity denoted the potential temperature remains unchanged. The potential temperature is equal to the temperature the parcel would have if it were raised or lowered through the atmosphere to a standard reference pressure, 100,000 Pa. The potential temperature plays the same role in the atmosphere—a compressible gas—that the usual temperature does in an incompressible fluid, like, to a good approximation, water.

3.3 Moist Thermodynamics

Humidity, the presence of water vapor in the air, has only a small quantitative influence on the thermodynamics of a rising parcel, except in the critically important case when saturation is reached and water vapor condenses. Then, the huge latent heat of vaporization of water comes into play, 2.5 million Joules for each kilogram of water condensed. To put this in perspective, the energy released in the atmosphere by condensing water vapor at the rate necessary to sustain a heavy rainfall of 1 cm/h is nearly 7000 W/m^2. This is more than five times the maximum rate at which the sun ever warms 1 m^2 of the earth's surface.

To deal with the thermodynamics of a parcel at saturation, it is typically assumed that condensation occurs as soon as the parcel is saturated and that the liquid water that condenses immediately falls out of the parcel as rain. Under the latter assumption, the thermodynamics become irreversible. The water that condenses is lost and cannot reevaporate into the parcel. Therefore, if a parcel rises past the altitude of its saturation, denoted the lifting condensation level, and then returns to its original altitude, it does not return to its initial temperature.

To incorporate condensation into this model, this model differs from the preceding one in two ways (Figure 3.5). First, the latent heat released when water vapor condenses in the parcel is a flow contributing to the Internal_energy stock. Second, the amount of water vapor in the parcel, represented by the mixing ratio, the ratio of the mass of vapor to that of dry air, must be represented by a stock. When the evaporation of rain or cloud water is neglected, the mixing ratio can only decrease by condensation. To

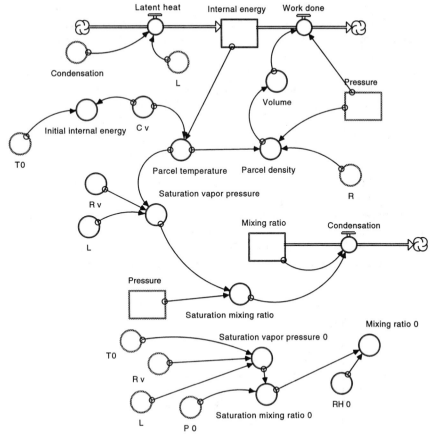

FIGURE 3.5   A model of the humidity and thermodynamics of a parcel.

compute the rate of condensation, an expression is needed for the concentration of water vapor at saturation, expressed as a mixing ratio. The Clausius–Clapeyron equation is a differential equation for the saturation vapor pressure derived from the thermodynamics of the Carnot cycle. Under the approximations that the volume occupied by liquid water can be neglected relative to that of water vapor, and that the latent heat of vaporization is independent of temperature, a useful approximate expression for the dependence of the saturation vapor pressure on temperature, T, may be obtained,

$$e\_s = 611 \ \{Pa\} \ exp[L/R\_v \times (1/T \ (0°C) - 1/T)]$$

This expression is used throughout this book in models that involve the evaporation and condensation of water (for example, Section 2.7). It yields an approximately exponential increase in saturation vapor pressure with temperature.

The mixing ratio of water in the rising parcel is conserved until the decreasing saturation mixing ratio matches the mixing ratio of the parcel, at which point a cloud appears and condensation of cloud droplets removes water vapor from the parcel. The saturation mixing ratio decreases as the parcel rises, because the temperature decreases; and the saturation vapor pressure, and thus the saturation mixing ratio, depend strongly on the parcel temperature. How high the parcel rises before saturating, the height of the lifting condensation level, depends on the humidity of the parcel when it started. The lifting condensation level for air at or near the surface of the earth provides a good estimate for the heights of the bases of convective (cumulus) clouds.

Figure 3.6 shows temperature profiles for rising air parcels with starting relative humidities at the ground of 0%, 50%, and 100%. The perfectly dry parcel never reaches its lifting condensation level, the 50% relative humidity parcel has its lifting condensation level at about 1200 m, and for the initially saturated parcel, the lifting condensation level is at the ground. Above the lifting condensation level, the latent heat of condensation contributes to

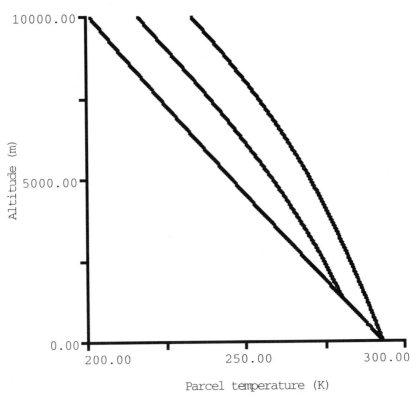

FIGURE 3.6   Temperature versus height for rising parcels with different initial humidities.

the internal energy of the parcel. Thus, it cools more slowly with height than does an unsaturated parcel. As will be discussed in Section 3.5, where we consider the dynamics of saturated parcels moving vertically in the atmosphere, this effect is responsible for most convective instability in the atmosphere, specifically, that responsible for thunderstorms.

As the parcel rises, less and less water vapor remains, so that at high altitudes the parcel again cools as rapidly with increasing altitude as one that was initially dry. Thus, as the altitude increases, the three curves in Figure 3.6 become nearly parallel, though an initially more humid parcel is warmer than a drier one at all altitudes.

3.4 Dry Parcel Dynamics

The dynamics of rising and falling parcels of air are influenced by their thermodynamics through their buoyancy. If a parcel is warmer than the air around it, it is, according to the ideal gas law, less dense, and it will accelerate upward. Including dynamics in our model of parcel thermodynamics requires only adding a stock for the vertical velocity, altered by a single flow for the upward buoyant acceleration (Figure 3.7), given by

$$g \times (\text{Parcel\_temperature} - \text{T\_profile})/\text{T\_profile}.$$

Because a parcel cools as it rises, if the temperature outside of the parcel (T_profile in the model and in the preceding expression) increases or decreases slowly with height, then a rising parcel is cooler than its environment. In this case, its buoyant acceleration is negative, and it accelerates back toward its starting height. The opposite is true for a falling parcel. The result is that the parcel oscillates about its initial altitude with a frequency, called the *buoyancy frequency,* that is lower when the environmental temperature decreases more rapidly with height. This effect is displayed in Figure 3.8, which shows the vertical motion of parcels in environments with lapse rates of 0, 2.5, 5, and 7.5 K/km. In all three cases, the parcel starts with an upward velocity of 0.1 m/s. The slower oscillations and

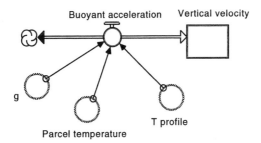

FIGURE 3.7   A model of the dynamics of a dry parcel.

larger-amplitude oscillations correspond to the greater environmental lapse rates. If the motion of the parcel is taken as analogous to that of a mass hanging from a spring, the buoyancy frequency may be thought of as the "spring constant" for vertical motions in the atmosphere. The smaller the temperature decrease or the more rapid the temperature increase with height, the "stiffer" the atmosphere is, and the more it resists the vertical displacement of air parcels. This is shown in Figure 3.8 by the increase in the maximum vertical displacements achieved as the lapse rate, Gamma, is increased. When temperatures increase, rather than decrease, with height, this is termed a *temperature inversion*. Temperature inversions near the ground inhibit vertical mixing and so can trap polluted air and lead to severe episodes of air pollution.

If the environmental lapse rate is sufficiently great—values of Gamma greater than the dry adiabatic lapse rate of about 1 K/100 m—although a rising parcel cools with height as before, after rising some distance, it is still warmer than its environment and so accelerates upward. In this case, there is a positive feedback between the parcel's displacement and its acceleration, and the atmosphere is unstable. The resulting vertical overturning is termed *dry convection*. In the atmosphere, dry convection generally occurs only in the lowest few meters above a surface strongly heated by the sun. In fact, one quality-control procedure applied to data from weather balloons is to discard data for which the lapse rate exceeds the critical value for this dry instability. In contrast, the occurrence of instability in the presence of moisture is an extremely important process in the atmosphere and is treated in the next model.

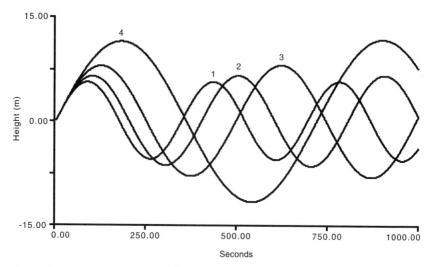

FIGURE 3.8   Vertical motion of dry parcels in environments with lapse rates of 0 (curve 1), 2.5 (curve 2), 5 (curve 3), and 7.5 K/km (curve 4).

3.5 Moist Parcel Dynamics

The moist parcel thermodynamic model is modified to include dynamics in the same way as was done for the dry parcel. Assuming that the relative humidity at the surface is not exactly zero, three outcomes are possible when parcels are launched upward from the surface. The vertical displacements in these three cases are shown in Figure 3.9. Here, the parcel starts at the surface with a relative humidity of 80%, and Gamma is set to 6.5 K/km. The parcel is launched from the surface with upward velocities of 4, 6, and 8 m/s. For the smallest initial velocity, the parcel never reaches its lifting condensation level, and it oscillates about its starting altitude just like a dry parcel; this is the case for curve 1 in Figure 3.9. No condensation occurs, and the mixing ratio remains constant. If the parcel crosses its lifting condensation level but is never warmer than its environment, then it falls back toward its starting level. In this case, however, the parcel has been irreversibly warmed by condensation, so it now oscillates about a new level slightly higher than its starting point; this can be seen in curve 2. The most interesting case—curve 3—is when the parcel rises to some altitude, called the *level of free convection,* at which it is warmer than its environment. Then the parcel accelerates upward.

The temperature of the parcel and its environment corresponding to curve 3 in Figure 3.9 are shown in Figure 3.10. Comparing these figures, it can be confirmed that the parcel accelerates upward whenever it is warmer than its surroundings. The upward acceleration ceases when the parcel again becomes cooler than its environment. This can occur, as in this model, when essentially all the vapor has condensed out of the parcel, so

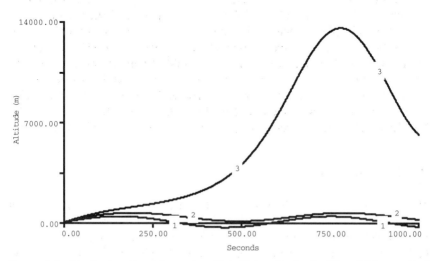

FIGURE 3.9   Vertical motion of parcels with initial vertical velocities of 4 (curve 1), 6 (curve 2), and 8 m/s (curve 3).

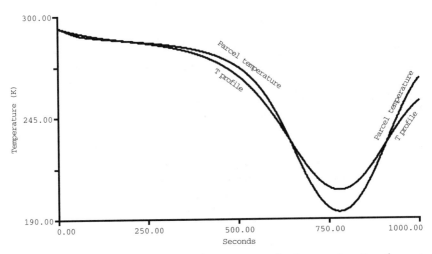

FIGURE 3.10   Parcel and environmental temperatures for the parcel motion shown in curve 3 of Figure 3.9.

that it cools with increasing height at the dry adiabatic lapse rate. In the atmosphere, this may occur when the parcel enters the stratosphere, where environmental temperatures cease their decrease with height. Through this process, a parcel of air that starts out near the surface ends in the upper troposphere. In the atmosphere, parcels do not overshoot as far as does curve 3 (Figure 3.9), because heat and momentum are mixed between the rising parcel and its environment, a process called *entrainment*.

The instability displayed by curve 3 (Figure 3.9) is denoted *conditional instability*. The temperature profile is stable for dry parcel motions, but instability occurs under the condition that the parcel is sufficiently moist when it leaves the surface. If the temperature decreases too slowly with height, however, the profile is again stable, as may be verified by varying the lapse rate, Gamma, in the model. Conditional instability gives rise to cumulonimbus clouds and thunderstorms. It is a key atmospheric mechanism for releasing latent heat and for maintaining the climatological temperature profile. Through conditional instability, latent heat provides the fuel for severe weather, such as squall lines, tornadoes, and hurricanes.

3.6 Mixing Clouds

The preceding models describe, among other things, clouds that form as parcels of air rise and cool by expansion. Clouds or fog can also form when air is cooled in some other way, either by coming into contact with a cold surface or by losing heat to space by infrared radiation. These two effects are common sources of fog. There is a third way in which clouds and fog form that is thoroughly ubiquitous but rather less easy to understand; this

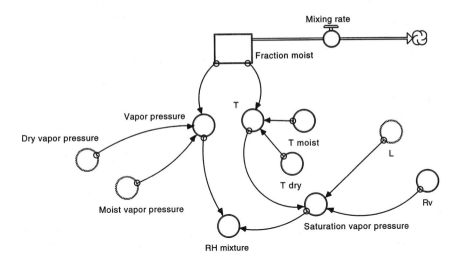

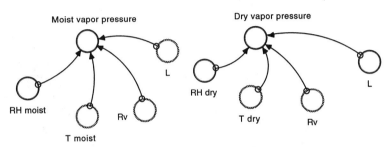

FIGURE 3.11   A model of mixing clouds.

is the formation of a cloud when air with different temperatures and humidities is mixed.

This model (Figure 3.11) treats the possible mixtures of air taken from two different sources, at different temperatures and humidities. The model is not really dynamical; time is used only as a "dummy" variable to consider all possible mixtures in turn. To fix ideas, it is helpful to consider a typical early winter situation in the central United States. The Great Lakes still retain much of their summer warmth and are not yet frozen. In the atmosphere, however, cold dry air has begun to appear over the lakes, brought from the Arctic by strong northwesterly winds. The cold, dry air arriving over the lake from the Northwest is one endpoint for a continuum of possible mixtures, and the warmer air in contact with the surface of the lake is the other. It is assumed that the lake air has the same temperature as the surface of the lake and is saturated at that temperature. Mixtures of air from these two sources are warmer than the Canadian air by an amount proportional to the difference in temperatures between the endpoints and also proportional to the fraction of lake-surface air in the mixture. The temperature, T, of the mixture is given by

$$T = \text{Fraction\_moist} \times (\text{T\_moist} - \text{T\_dry}) + \text{T\_dry}.$$

Similarly, the vapor pressure of the mixture is higher than that of the Canadian air by an amount proportional to the difference in the vapor pressures of the Canadian air and the lake air, and also proportional to the fraction of lake air in the mixture,

$$\text{RH\_mixture} = \text{Fraction\_moist} \times (\text{Moist\_vapor\_pressure}$$
$$- \text{Dry\_vapor\_pressure}) + \text{Dry\_vapor\_pressure}$$

The relationship between temperature and vapor pressure as the fractions of Canadian and lake air in the mixture are varied is shown by the straight line segments in Figure 3.12. Here, the Canadian air, found well above or upstream of the lake, is at the lower left end of each line segment; the lake

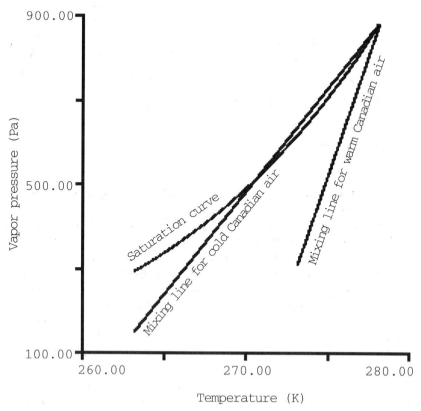

FIGURE 3.12   Mixing lines for warm and cold Canadian air passing over the Great Lakes. The saturation vapor pressure curve is shown for comparison.

air, the air in contact with the surface of the lake, is at the upper right end, and all possible mixtures lie on this line. The longer segment shows the case when the Canadian air is at −10°C, while the shorter segment corresponds to Canadian air at 0°C. In both cases, the lake is assumed to be at a temperature of 5°C. The saturation vapor pressure, however, has a nonlinear dependence on the temperature, as shown by the curve in Figure 3.12. For the case with warmer Canadian air, all possible mixtures of lake and Canadian air are unsaturated, and there is no cloud. For the case with colder Canadian air, however, the vapor pressure of the mixture exceeds the saturation vapor pressure for some mixtures, and a cloud forms. This cloud is called *steam fog,* and it can be seen wherever cold air blows over the surface of a warmer unfrozen lake, pond, puddle, river, or ocean. Figure 3.13 shows the relative humidity, the ratio of the vapor pressure to the saturation vapor pressure (× 100, so as to be expressed in percent) for both cases, as a function of the fraction of lake air in the mixture. For the colder Canadian air, the relative humidity, as expected, exceeds 100% for mixtures containing 40% or more lake air.

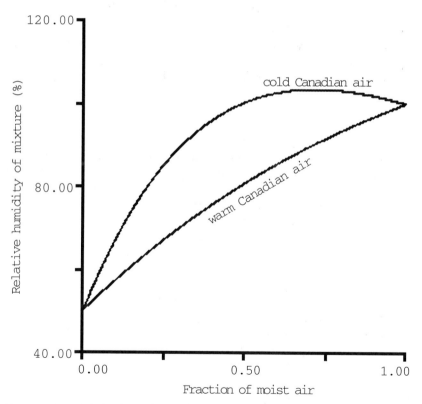

FIGURE 3.13   Relative humidities of mixtures of lake and Canadian air versus the fraction of lake air in the mixture.

Mixing clouds form in countless other places, among them, jet contrails and the visible cloud made by your breath on a cold day. One can explore these by simply changing the humidities and the temperatures of the end-points. Consider a steaming cup of tea sitting on your kitchen table. With this model, it is readily verified that there are three reasons that your cup may be steaming: your tea may be very hot, your kitchen may be very cold, or perhaps you have just taken a shower, and the air in your apartment is very humid.

Problems

3.1 The irreversible nature of the thermodynamics of a moist parcel of air can be exposed by considering what happens when humid air flows over a mountain range. Modify the model in Section 3.3 so that the parcel rises to a fixed altitude and then returns to sea level. To do this, the vertical velocity, W, must first be positive, then become negative after a set time. For example, inserting in the flow W,

IF (TIME < 1000) THEN 1 {m/s} ELSE −1 {m/s},

takes the parcel up to an altitude of 1 km and then back down. Note that the flows W, Work_done, and Pressure_change must be set as bi-flows to account for the descent of the parcel. How do the temperature and humidity in the lee of the mountain differ from conditions on the windward side? Where does the condensation (and thus the rain) occur?

3.2 When a parcel undergoes moist convective instability, the energy re-leased shows up in the kinetic energy of the parcel's vertical motion. The difference between the maximum kinetic energy of the parcel and its initial kinetic energy is a measure of the instability of the profile. This is called the *convective available potential energy (CAPE)*. The kinetic energy, per unit mass, is given by $.5 \times W^2$. Add a converter to the model in Section 3.5 to calculate the difference between the parcel kinetic energy and its initial value. Plot this kinetic energy difference, and read the maximum value off the graph to find the CAPE. Then, explore how CAPE varies when you vary the lapse rate, Gamma, the surface temperature, T0, and the surface relative humidity, RH_0. In practice, CAPE is determined from weather balloon observations and is used to estimate the likelihood of damaging thunderstorms. From your experiments, what conditions lend themselves to large values of CAPE and thus to severe weather? Why?

3.3 Use reasonable values of the parameters RH_moist, RH_dry, T_moist, and T_dry, in the mixing cloud model of Section 3.6, to confirm that a steaming teacup can indeed result from hot tea, a cold room, or high humidity. Here, the moist air is the air that is in contact with the surface of the tea, and the dry air is the ambient air in the room.

Further Reading

Comprehensive discussions of hydrostatic balance, atmospheric thermodynamics, and atmospheric stability can be found in *Atmospheric Science: An Introductory Survey,* by John M. Wallace and Peter V. Hobbs (1977, Academic Press, 467 pp.).

4

Dynamics of Horizontal Motion

4.1 Coriolis Deflections

The atmosphere and the ocean are fluids, and their dynamics are usually studied using the methods and equations of fluid dynamics. A great deal of understanding, however, can be gained by treating the motions of parcels of air as if they were solid bodies. The next model and its variations, then, consider the dynamics of a parcel of fluid moving horizontally on a rotating earth.

The dynamics of air or water flowing nearly horizontally in the earth's atmosphere or in an ocean involve three forces: those due to friction, horizontal variations in pressure, and the rotation of the earth. The last of these is called the *Coriolis force*. The Coriolis force is a so-called fictitious force, it exists only because we insist on referring motion to the rotating, and therefore accelerating, reference frame of the solid earth. Given, however, that it would be extremely inconvenient to do meteorology or oceanography in any other frame of reference, the Coriolis force has very real effects on the motions of the atmosphere and the ocean.

The dynamics in our model differ from familiar Newtonian particle dynamics only in the addition of the Coriolis force. This is a deflecting force, in that it acts at right angles to the velocity of a parcel. Its magnitude is proportional to the parcel speed. We define local Cartesian coordinates, x and y, where x increases to the east and y increases to the north. The associated components of the velocity are u and v, where u is the velocity in the x direction and v is the velocity in the y direction. Then the x and y components of the Coriolis force, per unit mass are

$$x\_Coriolis = f\,v$$
$$y\_Coriolis = -f\,u$$

where the Coriolis parameter, $f = 2\,\Omega\,\sin(\phi)$. Here, Ω is the angular frequency of the earth's rotation, 2π radians per day, or $7.29\ 10^{-5}\ s^{-1}$, and ϕ is the latitude. A typical midlatitude value for f is $10^{-4}\ 1/s$. The sine of the latitude changes sign across the equator, so the Coriolis force deflects motion

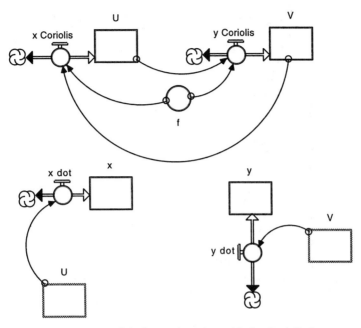

FIGURE 4.1   A model of parcel motion with the Coriolis force.

to the right in the Northern Hemisphere and to the left in the Southern Hemisphere.

If the motion is confined to a narrow range of latitudes, we can treat the Coriolis parameter as a constant. This is called the f-plane approximation, and it is equivalent to considering dynamics on a rotating turntable in the turntable's reference frame. Motions in the Northern Hemisphere are like those on a counterclockwise-rotating turntable, while those in the Southern Hemisphere are like those on a turntable rotating clockwise.

The simplest case is one in which only the Coriolis force acts on the motion of an otherwise freely moving parcel. The model (Figure 4.1) involves the interaction between the two components of the motion. The acceleration in the x direction is proportional to the velocity in the y direction, and vice versa. The trajectories of parcels with different initial speeds are compared in Figure 4.2 (note that the x and y dimensions of this graph are not the same). The parcels are launched moving to the east at initial speeds of 1, 10, and 100 m/s. The slowest parcel is most deflected from its initial eastward path, and the fastest parcel is deflected the least. This is a key feature of the Coriolis force; it acts most dramatically on objects that travel slowly over long distances, such as parcels of air in the atmosphere and parcels of water in the ocean.

If the same model is run longer (Figure 4.3), the parcel describes a circle. The parcels whose trajectories are shown in this figure are launched from

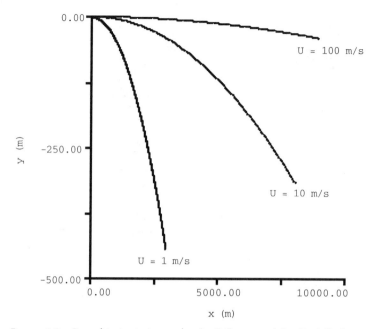

FIGURE 4.2   Parcel trajectories under the influence of the Coriolis force.

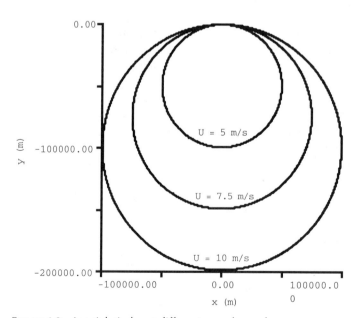

FIGURE 4.3   Inertial circles at different parcel speeds.

the origin with initial eastward velocities of 5, 7.5, and 10 m/s. If the initial speed is increased, the circle is bigger, but the time that is required for one orbit is unchanged. This may be confirmed by plotting the x or y position of the parcel against time for different initial velocities. These circles are called *inertial circles,* and the time for one orbit, $2\pi/f$, is called the *inertial period,* or the *pendulum day.* The latter refers to the time required for a Foucault pendulum to return to the same plane of motion in which it was started. The inertial period is long in the tropics, and it ranges from one day at 30° latitude to 12 hours at the poles. Motions with this period are prominent in the ocean but less so in the atmosphere.

4.2 Motion in Response to an External Force

Parcels of air do not move freely but are subject to external forces resulting from horizontal variations in the atmospheric pressure. The pressure typically varies by about 1000 Pa (10 mbar) over distances of 1000 km. Given that the density of air near the earth's surface is about 1 kg/m^3, this gives a force per unit mass, or a horizontal acceleration, of about 0.001 m/s^2. This value is 10,000 times smaller than the acceleration of gravity. Such small accelerations are important because they act in the horizontal, orthogonally to the force of gravity. The model in the preceding section is readily modified to include such an acceleration. For simplicity, a constant northward-directed force is chosen with a magnitude of 0.001 m/s^2, and for convenience, the units of time in this version of the model have been converted to days (86,400 seconds), and the units of length to kilometers.

It might be expected that a parcel starting from rest would be pushed northward by this imposed force. The parcel does start out moving northward, but it is soon deflected to the east by the Coriolis force, whereupon it undergoes a rapid eastward drift with small north–south oscillations superposed (Figure 4.4—note again the unequal scales of the x and y axes). Averaged over time, the y-directed Coriolis force is equal in magnitude and opposite in direction to the imposed force (Figure 4.5). The southward Coriolis force, a consequence of the eastward motion, balances the northward-imposed force. In the atmosphere and the ocean, this balance between the pressure-gradient force and the Coriolis force is ubiquitous and is denoted *geostrophic* (earth-turning) balance. Winds or currents in such balance are called *geostrophic winds or currents.* Except in the immediate vicinity of the equator, large-scale winds and currents are nearly geostrophic. Geostrophic motions are orthogonal to the pressure-gradient force, which is, in turn, orthogonal to the isobars (contours of constant pressure). Therefore, geostrophic flow follows the isobars. In the atmosphere, at levels well removed from the influence of surface drag, winds blow along the isobars, in the same sense as reproduced in this model, with high pressure to the right (in the Northern Hemisphere) when looking downwind.

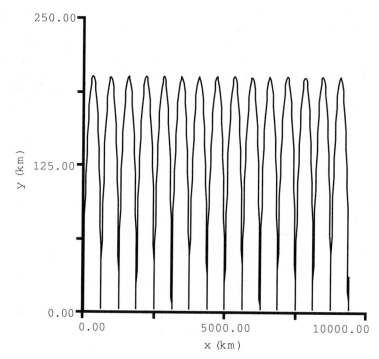

FIGURE 4.4   A parcel trajectory with a northward imposed force.

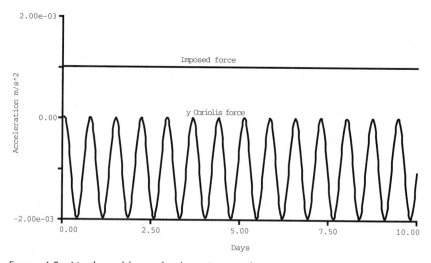

FIGURE 4.5   Northward forces for the trajectory shown in Figure 4.4.

4.3 Motions with Drag

The model in the preceding section displays a persistent oscillation with a frequency of f, the Coriolis parameter. Such oscillations are not generally observed in the atmosphere. They are removed primarily by the propagation of atmospheric waves, a process not readily simulated in STELLA, but they may also be damped out by friction. The inclusion of a simple linear drag in our model leads to a consideration of how atmospheric motions are modified by the influence of surface drag.

In this model (Figure 4.6), we begin with motion that is in geostrophic balance: The eastward motion of the parcel is such that its Coriolis deflection to the south exactly balances the imposed northward force. Then a linear drag is slowly turned on. The drag force is opposite in direction to the motion, and its strength is proportional to the speed with a proportionality constant denoted Drag_rate. Drag_rate is increased slowly over time, and we observe the locus of the velocity vector—the hodograph (Figure 4.7). The hodograph starts from motion that is purely eastward and geostrophic (U = 10 m/s, V = 0) but as the drag strengthens, the direction of the motion swings counterclockwise, and its speed weakens. For weak drag, the drag slows the motion in the *x* direction. This reduces the Coriolis

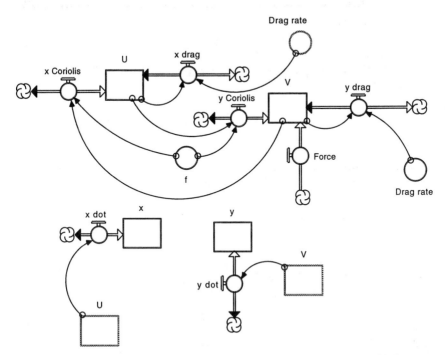

FIGURE 4.6   A model of parcel motion with an imposed force, the Coriolis force, and drag.

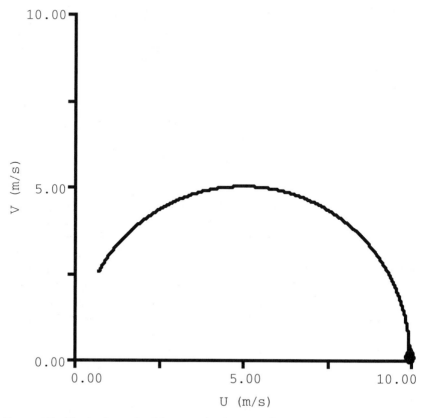

FIGURE 4.7   The hodograph of the parcel velocity as the Drag_rate is increased (right to left).

force and disrupts its balance with the imposed force, so that the parcel drifts northward, the direction of the external force. As the drag strengthens further, the motion swings more and more in the direction of the imposed force. Thus, in the atmosphere, winds in the boundary layer, that portion of the atmosphere subject to the influence of surface drag, blow across the isobars from high to low pressure. As the drag strengthens still further, the drag force becomes more important, relative to the Coriolis force, in balancing the imposed force. At the limit of very strong drag, the dynamics approach that of a "creeping" flow, wherein the imposed force is entirely balanced by drag, and the motion is in the direction of the imposed force. Observed surface winds in middle latitudes are typically deflected on the order of 45° from the along-isobar to the cross-isobar direction. This deflection varies with altitude above the surface, and a dynamical model of the vertical structure of the wind is provided in Section 4.8.

4.4 Motion Around Highs and Lows

The characteristic features of surface weather maps are high- and low-pressure systems, with their closed isobars. The pressure-gradient force then necessarily depends on the location, being directed either radially inward for a low or outward for a high. Motion around a high or low brings into play another fictitious force, the centrifugal force. Alternatively, uniform motion in a circle implies a constant acceleration toward the center of the circle that the Coriolis and pressure-gradient forces must provide.

In this version of the model (Figure 4.8), the components of the imposed force, Force_x and Force_y, are defined so as to point either radially inward

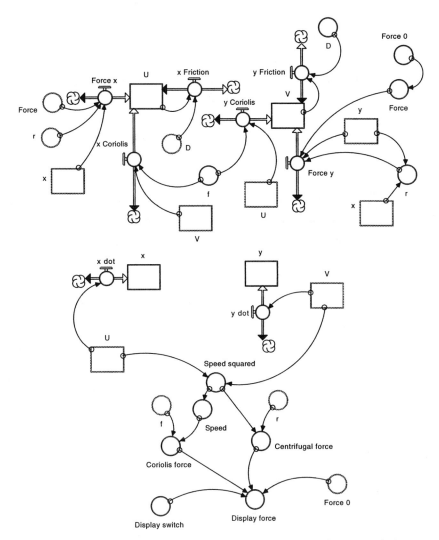

FIGURE 4.8   A model of parcel motion with an imposed force that acts radially into or out from the origin.

toward (Force < 0) or outward from (Force > 0) the origin ($x = y = 0$). The former corresponds to a low-pressure system and the latter to a high. The magnitude of the force is everywhere the same, and the parcel is launched at its geostrophic eastward velocity from a point north of the origin (this is eastward or positive for an outward force and westward or negative for an inward force). A weak drag is included to damp the initial inertial oscillations. Because drag is included, the parcel will slowly spiral either into or out from the origin, allowing a single run of the model to sample behavior over a range of radii.

For the low with its inward force, the speed of the parcel is always less than its geostrophic value, and the parcel slows as it spirals inward (Figure 4.9). Figure 4.10 shows the magnitudes of the three radial forces: the Coriolis force, the centrifugal force, and the (constant) imposed force.

(Here we digress once again to point out a STELLA modeling trick. It is sometimes desirable, as in the present case, to make a scatter plot with a few different variables on the same graph. This is accomplished by defining a new variable—here it is called Display_force—that with the use of IF-statements assumes the value of the different desired variables depending

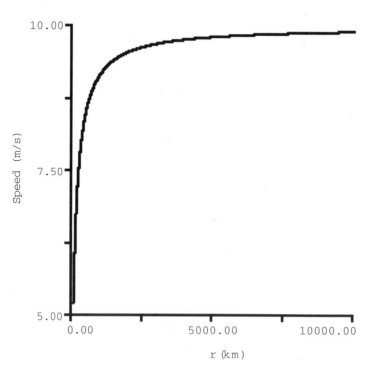

FIGURE 4.9   Parcel speed versus radius for a parcel spiraling into a low.

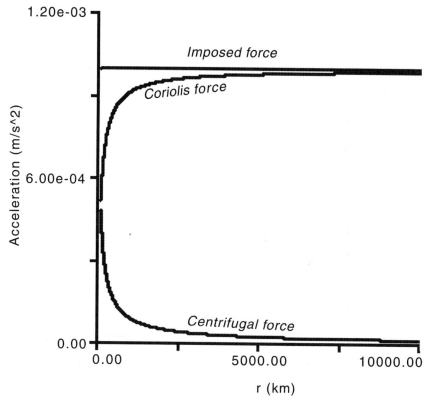

FIGURE 4.10   Radial forces versus radius for the parcel motion shown in Figure 4.9.

on the value of a switch variable (here denoted Display_switch). This latter variable serves as a sort of rotary switch that is stepped through its different values by making it the varying parameter in an S-Run. The different variables represented by the display variables can then appear on the same plot if the "Comparative" setting is selected.)

As time increases and the radius of the motion decreases, the centrifugal force becomes more and more important relative to the Coriolis force in balancing the inward imposed force. At very small radii, smaller than displayed in this figure, the centrifugal force can nearly balance the imposed force. This is called *cyclostrophic balance,* and it is found to be approximately valid in describing the flow around intense atmospheric vortices such as tornadoes and hurricanes. The speed around a low (Figure 4.9) is always slower than geostrophic, because both the centrifugal force and the Coriolis force push outward. A balance of radial forces can be achieved

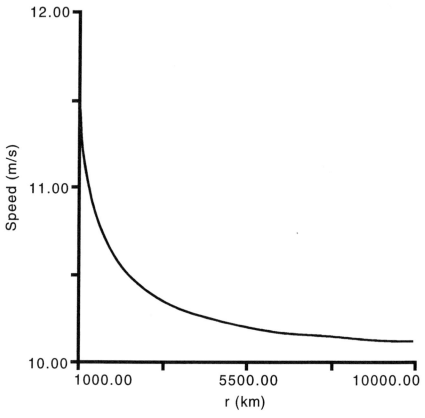

FIGURE 4.11   Parcel speed versus radius for a parcel spiraling outward from a high.

with a weaker Coriolis force and, therefore, at a slower speed than would be the case for linear motion.

For motion with an outward force (a "high"), the situation is exactly reversed. The centrifugal force works with the outward imposed force to oppose the Coriolis force, so that the Coriolis force must be larger than would be required to balance the imposed force alone, and the speed must be greater than geostrophic. Figure 4.11 shows the speed for a parcel spiraling slowly outward from a high, and Figure 4.12 shows the magnitudes of the radial forces. Given a fixed magnitude for the pressure gradient, no dynamical balance is possible close to the center of a high. The atmospheric consequence of this asymmetry between highs and lows is that low-pressure systems often have decreasing pressures right into their centers, while the pressure field is invariably flat near the center of a high.

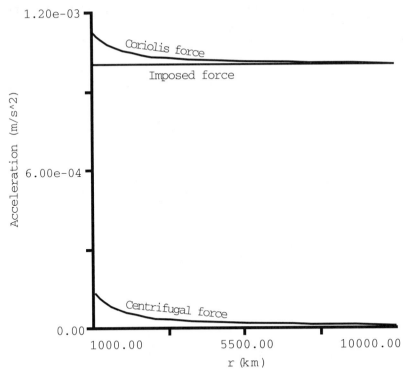

FIGURE 4.12   Radial forces versus radius for the parcel motion shown in Figure 4.11.

4.5 Motion on the Beta-Plane

The strength of the Coriolis deflection depends both on the speed of the parcel and on the magnitude of the Coriolis parameter. The latter is a function of the latitude. For many applications, it is useful to consider only the local linear variations of the Coriolis parameter about a central latitude. This is called the *beta-plane approximation*. The variation of the Coriolis parameter with latitude leads to some surprising parcel trajectories, especially in the vicinity of the equator. The "beta-effect" is added to the model of Section 4.1 by making the Coriolis parameter, f, depend linearly on y, the north–south position,

$$f = f0 + Beta \times y.$$

Here, Beta is the meridional derivative of the Coriolis parameter, Beta = $2 \, \Omega \, \cos(\phi)/a$, where a is the radius of the earth.

In middle latitudes, the beta-effect modifies parcel trajectories only slightly. Figure 4.13 shows the trajectory of a frictionless parcel released with an initial eastward velocity of 10 m/s from an initial position at 45°N. What were inertial circles in the absence of Beta here fail to close, because the rightward deflection of the motion is greater and the radius of curvature is reduced at higher latitudes. Because the motion is ever (in the Northern Hemisphere) turning to the right, the sharper turn at higher latitudes and gentler turn at lower latitudes lead to a gradual westward drift.

Near the equator, the effects of a varying Coriolis parameter are more dramatic. Because the sense of the Coriolis deflection reverses across the equator, a parcel that is generally headed eastward will always be deflected to the left—northward—when it is south of the equator, and to the

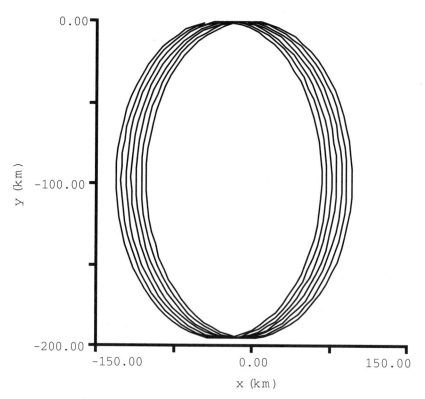

FIGURE 4.13   The trajectory of a parcel on the beta-plane, starting at 45°N.

right—southward—when it is north of the equator. Thus, the parcel follows the equator. Figure 4.14 shows as an example 10 days of the sinusoidal trajectory of a parcel released on the equator with an initial velocity of 10 m/s eastward and 1 m/s northward. If the parcel starts out moving to the west, however, and it has even a small component of motion to the north or south, it will be deflected further and further away from the equator, returning to the equator only after it has made a complete loop in higher latitudes. Figure 4.15 shows the trajectory, for 20 days, of a parcel started from the origin, again on the equator, with a velocity of 10 m/s westward and 1 m/s northward. Note that the maximum meridional displacements are more than 10 times larger than those shown in Figure 4.14.

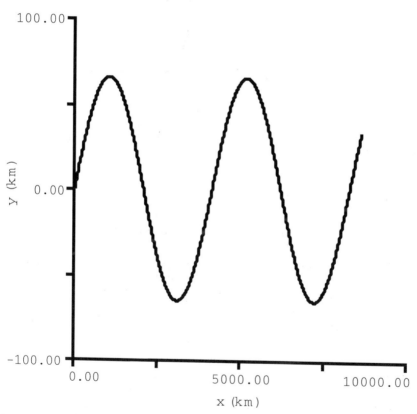

FIGURE 4.14   The trajectory of an eastward-moving parcel on the beta-plane near the equator.

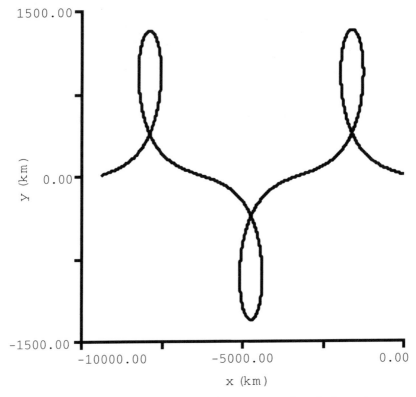

FIGURE 4.15   The trajectory of a westward-moving parcel on the beta-plane near the equator.

4.6 Inertial Instability

Earlier models in this chapter include an imposed force. When this force varies from place to place, the possibility arises of instability and, subsequently, nonlinear oscillations. Consider a parcel that starts from rest at a location where the external force vanishes. If the parcel starts with a small northward velocity, it is expected that its motion will, as shown in Section 4.1, describe an inertial circle. What if, however, there is a northward external force that increases in strength rapidly to the north? In this case, as the parcel moves northward and is deflected eastward by the Coriolis force, the external force may exceed the resulting southward Coriolis force. The parcel accelerates further northward and encounters an ever-stronger external force. This positive feedback loop is denoted *inertial instability*.

   In the atmosphere, the external force is due to the horizontal gradient of atmospheric pressure, and, because winds in the atmosphere are predominately geostrophic, it is customary to represent this force in terms of the associated velocity of the geostrophic wind. With this model, we focus on that region

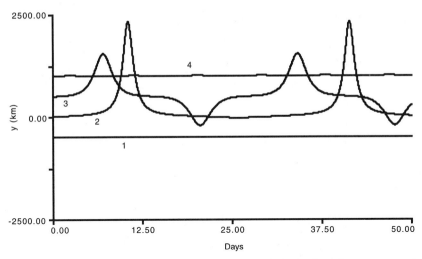

FIGURE 4.16   Parcel trajectories near the equator in the presence of cross-equatorial shear.

of the atmosphere where inertial instability is most likely to occur, near the equator. The elements of the model—the Coriolis force and an imposed force that vary with location—are exactly as before. We consider a west-to-east geostrophic wind that vanishes on the equator and linearly increases with latitude, so that the geostrophic winds are eastward north of the equator and westward south of the equator. The rate of increase with latitude is given by the parameter, Shear, in units of meters per second per kilometer. For the results shown below, a large value is used, 0.02 m/s/km. As in the earlier beta-plane example, the Coriolis parameter, f, increases linearly with latitude. Together, these conditions imply that the northward external force increases with the square of the distance from the equator. Parcels are launched at their local geostrophic speed in the x direction and with a small additional northward velocity, 0.1 m/s.

Figure 4.16 shows the meridional displacement of parcels so launched from positions 500 km south of the equator (curve 1), on the equator (curve 2), and 500 (curve 3) and 1000 km (curve 4) north of the equator. Parcels starting south or sufficiently far north of the equator undergo weak stable inertial oscillations around their geostrophic trajectories. For initial positions that are not too far north of the equator, however, the parcel rapidly accelerates away from its initial latitude. The net force in the y direction, the sum of the external and Coriolis forces, initially increases with the northward displacement of the parcel (Figure 4.17), confirming that this is indeed an instability. Once the parcel crosses a critical latitude, however, the net force reverses and pushes the parcel back toward the equator. The result is a nonlinear oscillation, in which parcels repeatedly traverse a wide range of latitudes.

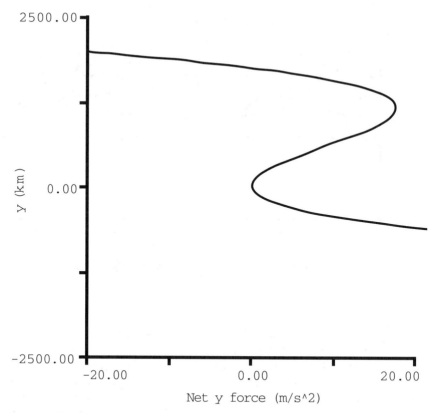

FIGURE 4.17   The net northward force in the presence of cross-equatorial shear on a parcel near the equator, versus its north–south position.

The atmospheric implication of this instability is that, when there is horizontal shear in the west-to-east winds across the equator, there is the potential for instability and a resulting rapid meridional mixing of the air in whichever hemisphere the winds are eastward. When the air is mixed, its momentum is mixed as well, and this mixing tends to eliminate the shear. Thus, we expect that through the process of inertial instability the atmosphere will tend to mix away cross-equatorial shear in the zonal wind.

4.7 The Ekman Layer

Having recognized that the Coriolis force deflects the motion of fluid parcels, it should not come as a surprise that when the wind blows over the ocean the resulting surface current is not simply downwind. This fact was made dramatically evident during the heroic days of Arctic exploration. Ships became frozen into the ice pack and would drift with the ice. It was observed that the ice and the ship drifted to the right of the direction of the

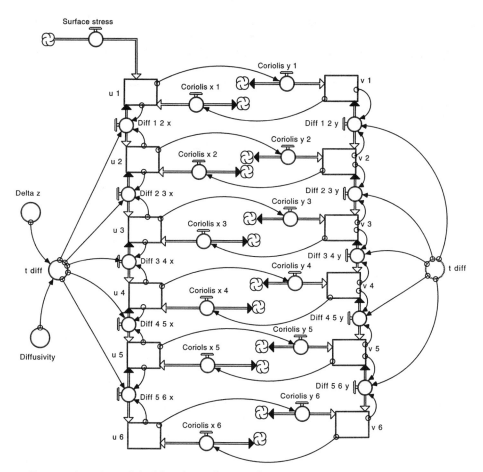

FIGURE 4.18   A model of the Ekman layer in the ocean.

wind. A physical argument to explain this observation was advanced by the Norwegian biologist and explorer Fridjof Nansen. Think of the ocean as comprising a stack of solid slabs. The slabs can slide independently, but there is friction between them. The wind pushes the top slab in the direction of the wind, and, because of friction between them, each slab pulls the one beneath it, and vice versa. In addition, the Coriolis force influences the motion of each slab. When the motion is steady, there is a three-force balance for each slab: the pull of the slab above (or, for the top slab, the wind), the Coriolis deflection to the right of the motion, and the retarding pull of the slab beneath. The result is that the surface current is to the right of the wind, and in each lower layer, the current is weaker and deflected further to the right.

A STELLA model of this process is readily constructed (Figure 4.18) that closely follows Nansen's physical description. The number of slabs is constrained only by the patience of the modeler; this version has six. The frictional interaction between neighboring slabs transmits, by the principle of action and reaction, as much momentum to the lower slab as it removes from the upper. Of course, except for sea ice, the ocean does not really consist of solid slabs. Rather, the frictional interaction of the slabs is a conceptually and computationally convenient model for the continuous diffusion of momentum vertically by turbulence. In this case, the ocean is treated as being infinitely deep, so there is no bottom friction on the lowest slab. For simplicity, the wind stress on the surface layer is assumed to be entirely to the east. It is turned on smoothly and approaches its final value asymptotically. The slabs of water are chosen to be 10 m thick, and the vertical diffusivity for momentum is set to a typical oceanic value of 0.01 m^2/s^{-1}.

The current vectors in the top five layers, after the model has reached a nearly steady state, are shown in Figure 4.19. The current decreases and

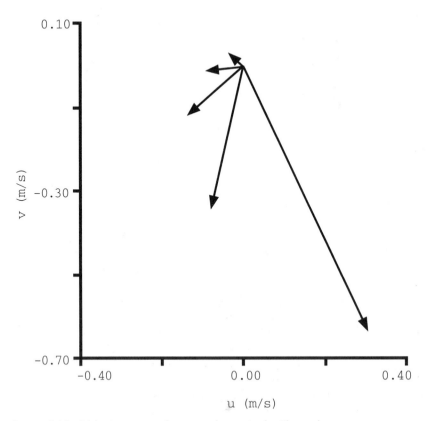

FIGURE 4.19   Velocity vectors for ocean layers in the Ekman layer.

rotates clockwise with depth, consistent with Nansen's argument. At many depths, the current has a westward component, opposite to the direction of the wind. A time series of any of the velocity components reveals a persistent oscillation that never damps out, even in a much longer simulation. These oscillations are at the inertial frequency, corresponding to a period of about 62,800 s. Because the only drag is *between* the slabs, there is no force to damp out the oscillations of all the slabs together. Such oscillations are present in the real ocean, though they are greatly reduced because the depth is much greater than 60 m. Also, any motion that extends through the entire depth of the ocean is damped by friction at the sea floor.

An important feature of the Ekman layer flow is the nature of depth-averaged motion. Neglecting the oscillations, the depth-averaged current is entirely at right angles to the wind (Figure 4.20)—in this case southward. This is as it must be, independent of the details of the interactions among the layers. The only net forces on the water are the wind stress and the Coriolis force. For these to balance, the depth-averaged current must flow 90° to the right of the wind. This motion is denoted Ekman drift, and it is important throughout the ocean. For example, surface winds typically blow southward along California's Pacific Coast. The Ekman drift carries surface water to the right of the wind, in this case westward, away from the coast. The surface water that is so removed is replaced by water upwelling from greater depths. The upwelling water is both colder and contains more biological nutrients than the water it replaces. The nutrients make for a biologically rich marine environment, but the cold makes for frequent fog and unpleasant swimming.

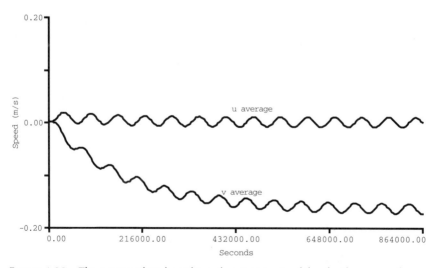

FIGURE 4.20   The eastward and northward components of the depth-averaged velocity in the Ekman layer.

4.8 The Ekman Layer in the Atmosphere

The atmosphere near the ground is closely analogous to the upper ocean. The ground exerts drag on the lowest layer of the air, which then, by the turbulent exchange of momentum, exerts a drag on higher layers, and so on. In the atmosphere, a horizontal pressure gradient is usually present, and it is expected that the winds are approximately in geostrophic balance at altitudes well removed from the surface. Horizontal variations in the atmospheric pressure vary only slowly with height, so it is reasonable to assume that the same horizontal pressure gradient force acts at all levels in the boundary layer.

With these considerations, we can establish a STELLA model of the atmospheric Ekman layer. The present model is identical to the previous one with a few exceptions:

1. There is no wind stress exerted at the top of the model.
2. There is at every level an identical horizontal pressure-gradient force. For simplicity, this is chosen to act in the y direction, yielding geostrophic motion in the positive x direction.
3. The lowest level of the atmosphere exchanges momentum with the ground.

The values of numerical constants are also changed. The layers are now 200 m thick, and the diffusivity is 5 m²/s. The pressure gradient, chosen to yield a geostrophic wind of 10 m/s, is turned on gradually over 2 hours. The results (Figure 4.21) differ from the marine case in two respects. First, the winds at the top of the Ekman layer, furthest from the ground, are nearly geostrophic. Second, the winds at lower layers are deflected to the *left* relative to the geostrophic wind, with the wind vectors at all levels lying in or near the first quadrant. This difference may be understood by imagining that one is at rest relative to the geostrophic wind. In a reference frame at rest with the geostrophic wind, it appears that the ground is being dragged beneath the atmosphere in a direction opposite to the geostrophic wind. This dragging of the ground along the bottom of the atmosphere is analogous to the wind stress on the surface of the ocean, so that the atmospheric Ekman layer is like the oceanic Ekman layer turned upside down. Confirmation that the winds in this model are in the right direction comes from a comparison with the results obtained for particle motion in the presence of friction (Section 4.3). That model showed that the winds in the boundary layer must blow to the left of geostrophic winds, consistent with the present model, because by so doing, they have a component of motion in the direction of the pressure-gradient force.

Problems

4.1 The model for parcel motion on a beta-plane, Section 4.5, can, with some effort, be modified to account for motion on a spherical earth. First, the spatial coordinates of the parcel, x and y, must be replaced

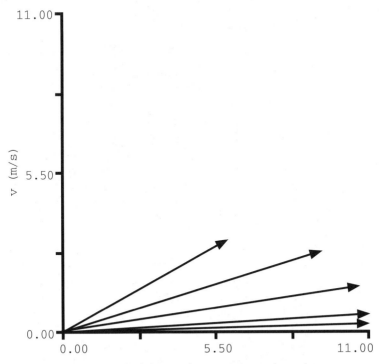

FigURE 4.21   Velocity vectors for layers in the atmospheric Ekman layer.

by longitude (Lon) and latitude (Lat). The flows x_dot and y_dot then become

$$x\_dot = U/(a \times COS(Lat)), \quad y\_dot = V/a$$

where a is the radius of the earth, 6.37e6 m.

The more complicated modifications come about because the surface of the earth is curved, and because centrifugal, as well as Coriolis, forces are associated with motion on the surface of a spherical planet. The flows representing the Coriolis force become

$$x\_coriolis = V \times (f + U/a \times TAN(Lat)) \times 86{,}400 \ \{s/day\},$$
$$y\_coriolis = -U \times (f + U/a \times TAN(Lat)) \times 86{,}400 \ \{s/day\}.$$

Here the Coriolis parameter, f, is given by

$$f = 2 \times Omega \times SIN(Lat),$$

where Omega is the angular frequency of the earth's rotation, 7.29e–5 {1/s}. The numerical constant 86,400 is needed to express time in days rather than seconds. Making these modifications requires additional connections among the stocks, flows, and converters. Lat and Lon are given in units of radians. To express them in more familiar units, use a converter that multiplies them by the factor, 180/PI. To keep the longitude within its conventional bounds, use

$$Lon\_degrees = MOD(Lon \times 180/PI/360) - 180$$

Perhaps the most instructive experiment that can be done with this model is to consider the motion on a nonrotating planet, by setting Omega to zero. What do you expect will happen to a parcel launched due east or west? Can you explain your results?

Similar results will be obtained with Omega returned to its usual value if very large initial velocities are used (remember to reduce the timestep). Why?

Finally, you will find that this model produces nonsensical results if a parcel is launched from either pole, or if its trajectory takes it over the pole. Why?

4.2 In the model for inertial instability (Section 4.6), experiment with modifying the value of the cross-equatorial wind shear, given by the parameter, Shear. How does the meridional excursion or the parcel vary with this parameter? Can you achieve stability for a parcel on the equator for any value of Shear, however small?

4.3 Modify the oceanic Ekman layer model, Section 4.7, to include bottom friction. To do this, add diffusive transports of momentum out of the stocks, u_6 and v_6. These should resemble the diffusive flows between the model layers, except that they end in clouds. How does bottom friction modify the Ekman layer? Reduce the depth of the entire model by decreasing Delta_z (a much shorter timestep will be required). What do the results tell you about the vertically integrated, wind-driven transport in a shallow body of water, in comparison with that in a deep one?

Further Reading

The standard introduction to atmospheric forces and balanced motion, as well as to the Ekman layer is *An Introduction to Dynamic Meteorology (Third Edition),* by James R. Holton (1992, Academic Press, 511 pp.). A complete discussion of the oceanic Ekman layer may be found in *Introduction to Geophysical Fluid Dynamics,* by Benoit Cushman-Roisin (1994, Prentice Hall, 320 pp.).

5

Dynamics of Circulation and Vorticity

5.1 Simple Convection

In the preceding two chapters, we looked at the dynamics of the atmosphere in terms of the motion of fluid parcels, discrete bits of fluid that obey Newton's laws of motion. Large-scale motions in the atmosphere and ocean—and, in fact, persistent motions in any fluid—are typically characterized by flow around closed paths. The strength of the flow around a closed path, denoted the *circulation,* is loosely analogous to the angular momentum of a solid body. More precisely, the circulation is the integral of the along-path component of the fluid velocity around a closed path. Just as the angular momentum is altered by torques, the circulation is altered by quantities akin to torques. The models in this chapter range from convection—the vertical overturning of a fluid heated from below—to the Gulf Stream—the western boundary current in the North Atlantic Ocean. They are unified by the fact that the dynamics in all cases are treated in terms of sources and sinks of the circulation, or its local manifestation, the vorticity.

In Chapter 3, we considered the stability of a parcel of air, and we found that, for a dry parcel, the motion was unstable if the decrease in temperature with height exceeded some critical value. Including the circulation permits a more fluid-dynamical view of the process of convection. Here, we model a fluid circuit that starts off with cooler, denser, fluid above and warmer, less dense fluid, below. This situation is shown schematically in Figure 5.1. It is closely analogous to a bicycle wheel with a weight attached to one spoke near the rim, and starting with the weight directly above the hub. When the wheel is given a small push, the weight is moved to one side, and gravity then provides a net torque that increases the rotation of the wheel, moving the weight further off center, increasing the torque, and thus the acceleration, and so on. This is a positive feedback on the motion, and it yields instability. A fluid that starts with warmer fluid below and cooler above is similarly unstable to overturning motions.

Our model (Figure 5.2) describes the circulation around a square circuit where each side is of length H. The temperatures at the top, bottom, left, and right of the fluid circuit are treated by the loop at the top of the model.

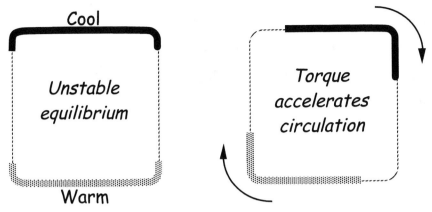

FIGURE 5.1   A schematic of convection.

A single stock represents the temperature on each side of the circuit. These temperatures are altered only by advection of warm or cool fluid from one stock to the next. The temperatures influence the flow through the gravitational torques on this vertically oriented fluid circuit. Circulation, the lower stock, is driven by the flow denoted Solenoidal—the sum of the gravitational torques around the circuit. Counterclockwise circulation is defined as positive. The sum of gravitational torques is, in turn, proportional to the difference in the density between the right and left sides of the circuit and, therefore, the difference in the temperatures. Once a circulation starts, these temperatures are changed by advection.

When this model is run without drag, and starting with a very small counterclockwise circulation, the motion accelerates as warm, positively buoyant fluid arrives on the right from below and as cool, negatively buoyant fluid arrives on the left from above. This increases the gravitational torque, which accelerates the circulation, and further enhances the temperature contrast. When fluid from the top and bottom have completely replaced the fluid on the left and right, respectively, the torque is at a maximum. Thereafter, the temperature contrast, Delta_T, between the left and right diminishes, though the torque is still positive, and the circulation continues to increase. Finally, the cold fluid occupies the bottom, the warm fluid occupies the top, the sides are at equal temperatures, and the torque vanishes. The remaining momentum in the circuit is sufficient to return the cold fluid back to the top, and the warm fluid to the bottom. In this stage, the torque opposes the circulation, and it slows. When the cool fluid is back at the top, just the initial momentum of the circulation remains, to carry it over the top and repeat the cycle. Figure 5.3 shows the circulation and the temperature at the bottom. When the bottom is cool, the circulation is large. When the bottom is warm, the circulation is weakest. The system is briefly poised near an unstable equilibrium before the warm fluid again rises to the top.

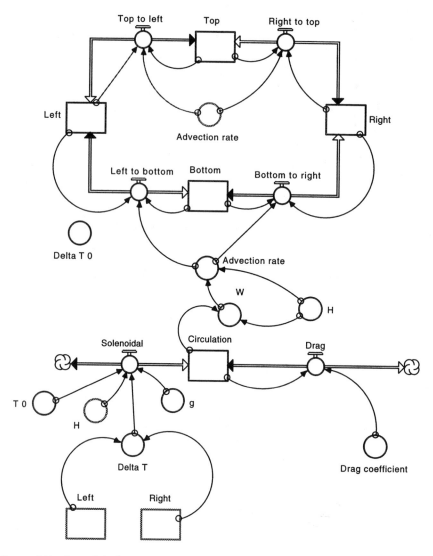

FIGURE 5.2   A model of convection.

The behavior changes when drag is included. Figure 5.4 shows the temperature in all four stocks when the model is run with a damping time of 20 seconds on the circulation. Now the circulation is slowed before it can bring the cool fluid all the way back to the top; it remains slightly to the right when the circulation comes to a stop, so that the motion reverses direction. After a few such damped oscillations, the fluid comes to rest in its stable configuration, with cool fluid at the bottom and warm fluid above.

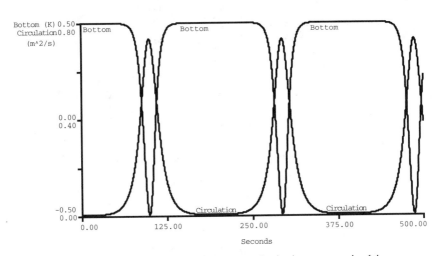

FIGURE 5.3   The circulation and the temperature in the bottom stock of the convection model.

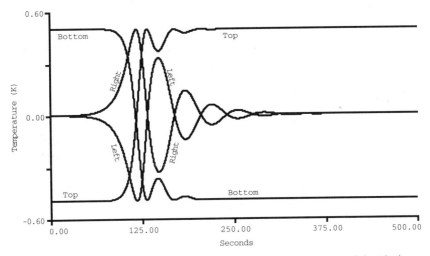

FIGURE 5.4   The temperature in all four stocks of the convection model with drag.

5.2 Convection with Conduction

In almost any real fluid, either conduction or turbulent mixing moves heat from warm to cool regions. Also, convection generally occurs when a fluid is persistently heated from below and/or cooled from above. This next model (Figure 5.5) includes these effects. There are now sources and sinks of heat at the bottom and top of the fluid, and heat is transferred diffusively as well as advectively between the top and bottom, and between the left

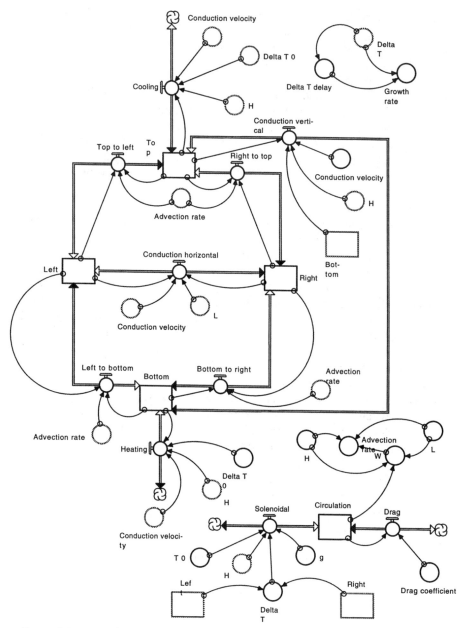

FIGURE 5.5   A model of convection including heat transfer by conduction.

and right sides, of the circuit. A careful examination of this model reveals that beyond these changes just noted a different method is used to calculate the advection of heat between the stocks. The previous model used a centered differencing scheme, in which the temperature of the fluid flowing between any two stocks was assumed to be the average of the temperatures of the same two stocks. In a model of this sort, a centered differencing scheme leads to a model that rigorously conserves energy, as was clearly evident in the case with no drag. In the present case, however, in which thermal forcing is included, the centered differencing leads to unphysical behavior, resulting from the fact that with centered differencing the net advective flow of heat into or out of a stock does not depend on the temperature of that stock, only on its neighbors. Instead, here, an upstream differencing scheme is employed, in which the fluid leaving a stock is assumed to have the temperature of that stock. This is the same scheme shown in Section 2.5. Although all numerical models are sensitive to the numerical schemes employed, the extreme sensitivity in this case is a consequence of the small number of stocks being used, or equivalently, the very coarse resolution of the model.

Convection develops as a positive feedback between the circulation and the difference in the temperature and thus density between the rising and sinking regions of the fluid. In its early stages, before the warm and cold fluid have had a chance to travel very far from the lower and upper boundaries, the growth of the circulation is exponential. This is an example of a linear instability. If an unstable system, such as this one, is perturbed randomly so that there are initially small motions on many different length scales, it might be expected that those motions that grow most rapidly are the ones that will come to dominate the behavior and be most clearly visible to the observer. Thus, it is of interest to find out the scale of the fastest-growing solution. In our STELLA model, we have selected the spatial dimension of the convective circuit, while in an unbounded fluid in nature, the fluid dynamics pick the dimensions. At least initially, these will be the dimensions of motions that grow most rapidly. For an exponentially growing solution, the strength of the perturbations—here we use the difference in the temperature between the rising and sinking branches of the flow—is proportional to $\exp(g\,t)$, where g is the growth rate. Equivalently, the growth rate is proportional to the logarithm of the ratio of the perturbation amplitudes at two different times.

As an example, we consider how the growth rate for convection in the present model depends on the width of the convection cell and the thermal conductivity, represented here as a conduction velocity. The height of the cell, H, is fixed at 1 m, while the width of the cell, L, is varied in small increments from 0 to 2 m, and four different values, 0.05, 0.1, 0.15, and 0.2 m/s, are used for the conduction velocity. Figure 5.6 shows the results. As might be expected, when the conduction velocity is largest (the bottom curve), the convection grows most slowly. For many widths, there is no growth at all, because the conduction of heat between the rising and sinking fluid dissipates the temperature difference faster than it can be generated by advection. As the conduction velocity is reduced, the growth rate increases, but the width at which the fastest growth occurs decreases. Reducing the heat conduction allows the rising and sinking fluid to be

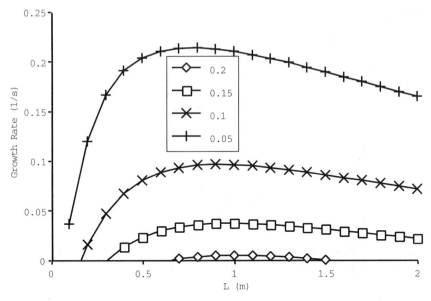

FIGURE 5.6   Convective growth rate versus cell width for different thermal conductivities.

closer together without their temperature contrast being destroyed. This favors a narrow width, or tall, narrow convection cells, since the mechanical drag on the circulation in this model is proportional only to the circumference of the cell. If there were no conduction, the fastest growth would be a cell with the smallest possible circumference, one that was infinitely narrow.

Cellular convection is often present in the atmosphere, most frequently in winter when cold air spills from a continent over a warm ocean or lake. The convection can form closed cells or long rolls. The clouds that form in the rising air make the structure of the convection visible in satellite images. It is very difficult to predict the horizontal scale of the convection cells, because in the atmosphere the heat conduction and friction are not simply parameters but are determined by turbulent mixing that itself depends on the nature of the convective motion.

5.3 Vorticity Dynamics—the Vortex Pair

The circulation, the tendency of a fluid to flow about a closed curve, is a macroscopic property of a fluid flow. The vorticity is the local property that relates to the circulation. The vorticity is the circulation divided by the area enclosed by the circuit in the limit in which this area is taken to zero. More simply, the vorticity is the local tendency for a fluid to rotate. While a fluid can rotate around an axis oriented in any direction, here we consider flows confined to a horizontal plane, so that rotation is only about a vertical axis. This is a reasonable metaphor for the earth's atmosphere or ocean, where the large-scale flow is quasi-horizontal. In this case, the vorticity can be visualized as the rate of rotation of a paddle wheel placed in the fluid with its axis vertical. While more abstract than the circulation, the vorticity is often more useful, because it is conserved under many circumstances of practical relevance. In other words, a bit of fluid retains the same rate of rotation as it moves.

Generally, because the vorticity is a property of the fluid at every point, it is not readily exploited in STELLA models. There is, however, a special case that is amenable to STELLA modeling, the point vortex. In a point vortex, the rotation is confined within a region of vanishing area. Right at the vortex, our paddle wheel would spin infinitely fast, while elsewhere it would not spin at all. While this is clearly unrealistic, it is true that in many real fluid flows the vorticity is concentrated in a small region. Consider, for example, a strong atmospheric cyclone. For present purposes, the key point is that when the dynamics of a fluid are treated using point vortices, they become the dynamics of a finite number of interacting entities and can easily be represented in STELLA.

The point vortex is surrounded by a circulation that is the same at all separations from the vortex. Imagine that the vortex is in the center of a circle. The circumference of this circle is proportional to its radius. If the circulation, which is simply the tangential velocity multiplied by the circumference, is the same at all radii, then the fluid velocity induced by the point vortex must vary inversely with the radius. The velocity of the flow at any point in a fluid can be constructed by simply adding the contributions from all the point vortices in the fluid. These vortices are, in turn, advected by the flow, and the system is closed.

A system with a single vortex is not interesting. The fluid flows around the vortex, but the vortex does not move. Once there are two vortices, they can interact, with each vortex carried along by the flow induced by the other. Since the flow induced by each vortex is tangential to a circle surrounding the vortex, there is no radial motion, and so the vortex separation never changes. The model of a two-vortex system is shown in Figure 5.7. If the vortices have the same sign and magnitude, they simply revolve around a point midway between them. Figure 5.8 shows the trajectories of two such vortices. If one of the vortices is stronger than the other, the two revolve about a point closer to the stronger vortex. An example is shown in Figure 5.9. Here, vortex 2 is twice as strong as vortex 1; both vortices start on the y axis with vortex 1 at $y = 1$ and vortex 2 at $y = -1$. Vortex 1 describes a circle with a radius twice that described by vortex 2, and they revolve around a point that is closer to the initial position of vortex 2. In the limiting case where one vortex is much stronger than the other, the

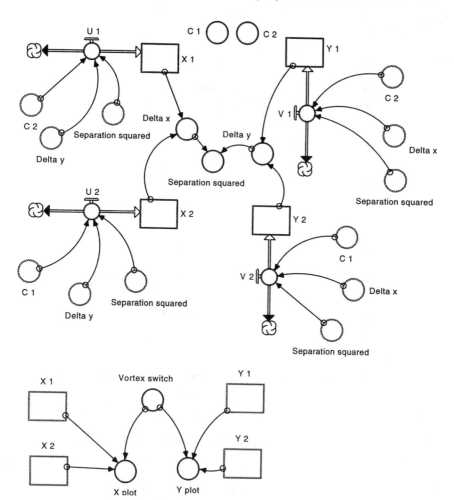

FIGURE 5.7   A two-vortex model.

stronger vortex barely moves while the weaker one revolves around it, much like a light satellite in orbit about a massive planet. This analogy to planetary orbits fails, however, when we consider vortices with different signs. When the two vortices are equal in magnitude but opposite in sign, each advects the other in the same direction at exactly the same rate. Such a vortex pair, therefore, "swims" in a straight line (Figure 5.10). The figure shows the case in which the upper vortex is positive with counterclockwise circulation, and the lower vortex is negative with clockwise circulation. Each vortex moves the other to the right.

Such vortex pairs provide a conceptual model of a meteorological phenomenon called *blocking*. In a block, the atmospheric circulation over

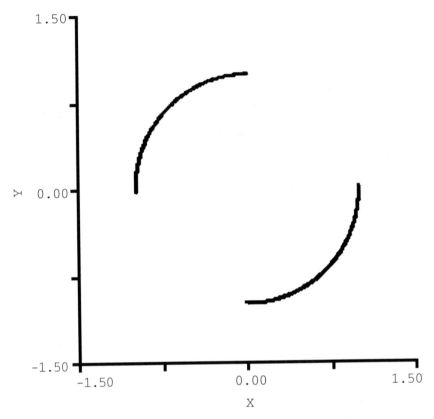

FIGURE 5.8   Trajectories of two point vortices with the same sign and equal strengths.

some region, often the North Atlantic Ocean, seems to be frozen, often for a week or more. Blocks often have an anticyclone positioned directly north of a cyclone, i.e., a negative vortex north of a positive one. Considered as a vortex pair, this pattern should translate to the west, but because the mean winds are from the west, the pair can remain in the same place. The block holds its position by swimming against the current.

If the vortices are of opposite sign but unequal strength, the vortices tend to move each other in the same direction, but the stronger vortex moves the weaker one more, so that the vortex pair describes a large circle with the weaker vortex on the outside. This is something like sculling with two rowers of unequal strength. Figure 5.11 shows the trajectories for the same case as Figure 5.10, only now the negative, initially lower, vortex is half again as strong as the positive one. The vortex pair follows a clockwise trajectory about a large circle with the negative vortex on the inside.

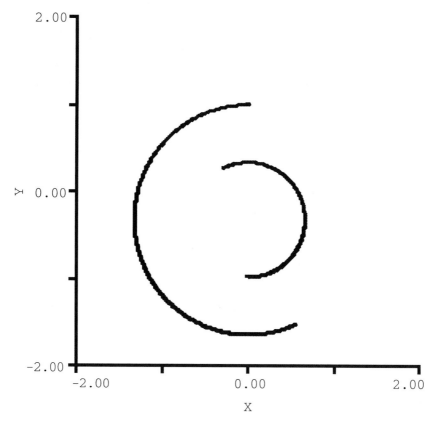

FIGURE 5.9   Trajectories of two point vortices with the same sign and unequal strengths.

5.4 Multiple Vortices

What happens when there are more than two vortices? Of course, it is expected that the behavior becomes more complex. For two vortices, there is just one interacting vortex pair, for three vortices there are three such pairs, and for four vortices there are six. In general, for N vortices, there are $N(N-1)/2$ interactions. Since each vortex is also a particle that is advected by the flow, some insight into the dynamics can be gained by considering the trajectory of such a particle. The multiple-vortex model shown here (Figure 5.12) allows for four vortices, but we begin by setting the circulation of two of them, C_3 and C_4, equal to zero. For simplicity, we consider the case where the remaining two vortices have the same sign and strength. We know from before that vortices 1 and 2 then describe circles.

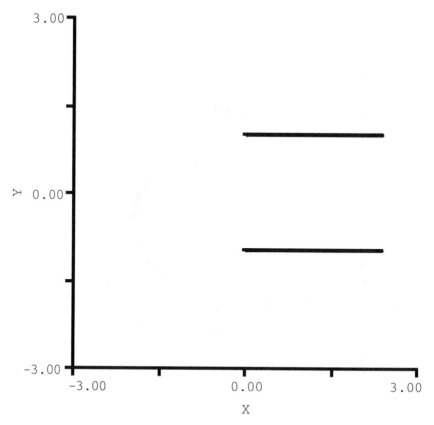

FIGURE 5.10   Trajectories of two point vortices with opposite signs and equal strengths.

Figure 5.13 shows the trajectory of a particle in the time-varying flow produced by the revolving vortices depicted in Figure 5.8. The trajectory of this passive particle is regular but complicated; its orbit rotates with the revolving vortex pair. When C_3 is set to the same sign and strength as C_1 and C_2, its motion now affects vortices 1 and 2, so that their motion is more complicated (Figure 5.14), though still regular. Most surprising is the trajectory of the remaining passive particle, started from the point $X = 0$, $Y = 0.1$, which is now very irregular (Figure 5.15). It is, in fact, chaotic. That the motion of a particle carried in a complicated but regular flow can be chaotic is significant for the redistribution of trace gases, such as water vapor, in the atmosphere (see Section 6.4). Finally, when this fourth particle is given a nonzero vorticity, its irregular motion is transmitted to the other vortices, and all four move chaotically. The resulting chaotic trajectory of vortex 1 is shown in Figure 5.16. In the atmosphere, there are many more than four centers of vorticity. From the perspective of the multiple-vortex

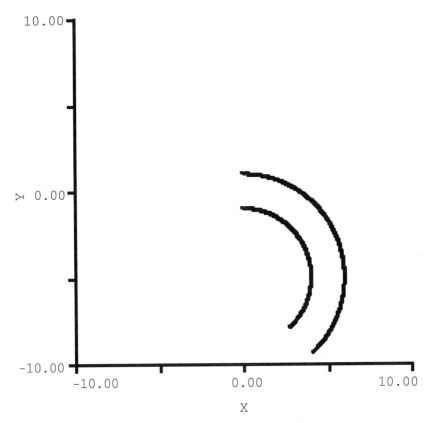

FIGURE 5.11   Trajectories of two point vortices with opposite signs and unequal strengths.

problem, it is inevitable that the motion in the atmosphere is chaotic and, therefore, inherently limited in its predictability. Here the chaos is in a system with no forcing or dissipation. Such a system that perfectly conserves its energy is denoted *Hamiltonian*. Sections 6.2 and 6.3 introduce chaotic, and hence unpredictable, systems in which energy is supplied by external forcing and is removed by dissipation—forced dissipative chaotic systems.

   While the general behavior of a set of four point vortices is chaotic, in many cases, the symmetry in the arrangement of the vortices prevents chaotic behavior. One such case arises from the use of image vortices to represent the presence of a solid boundary. When a fluid flow is represented by a collection of point vortices, the condition that fluid cannot flow across a linear solid boundary is satisfied by assuming a symmetric distribution of vortices centered on the boundary. Consider the case where the boundary is the y axis, and the fluid occupies the left half-plane. If two

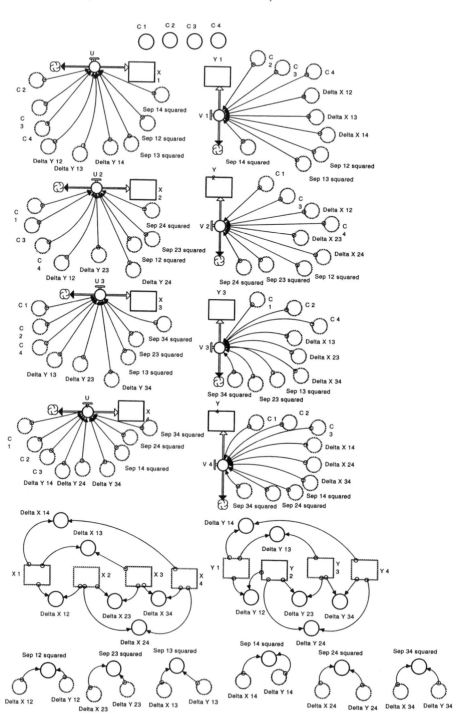

FIGURE 5.12   A multiple-vortex model.

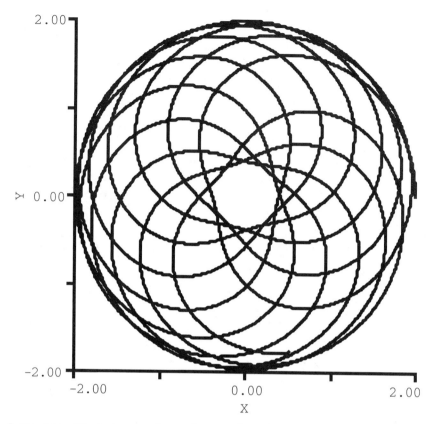

FIGURE 5.13   The trajectory of a passive particle in the flow generated by two equal vortices.

vortices are in the physical domain, then the existence of the boundary can be modeled by adding two additional image vortices positioned with the same strength and y coordinates as the real ones and with complementary x coordinates.

Figure 5.17 shows an example. A vortex pair is in the physical domain, the left half-plane, comprising equal and opposite vortices with a positive vortex starting at $(-1.5,0)$ and a negative vortex starting at $(-1,0)$. Unlike the previous case of a vortex pair (Figure 5.10) that "swims" in a straight line, the interaction of this pair with the image vortices in the positive half-plane causes it to turn toward the wall. This result suggests a simple explanation for the common observation that rising plumes of air tend to attach themselves to vertical or sloping boundaries. A plume of air that is rising straight up has regions of equal and opposite vorticity on opposite sides of the plume's center. If the plume is close to a wall, it will be deflected toward the wall by interactions with the image vorticity on the other side, as

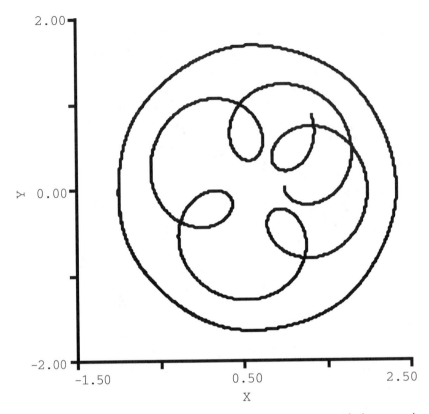

FIGURE 5.14   The trajectories of two of the vortices in a system with three equal vortices.

shown in Figure 5.17. Once the plume is in contact with the wall, however, the destruction of vorticity in the viscous or turbulent boundary layer adjacent to the wall must be taken into account.

5.5 The Rossby Wave

The interactions of point vortices have the character of action at a distance. This makes modeling fluid flows with vorticity problematical for STELLA. It is easy to imagine a STELLA model with a large two-dimensional array of stocks, each connected by flows to its immediate neighbors. If, however, each stock influences and is influenced by every other stock in the array, the model rapidly becomes a mess. Therefore, STELLA is not well suited for modeling a fluid flow with a general spatial distribution of vorticity. Much can be learned, however, from considering a simplified problem, a fluid

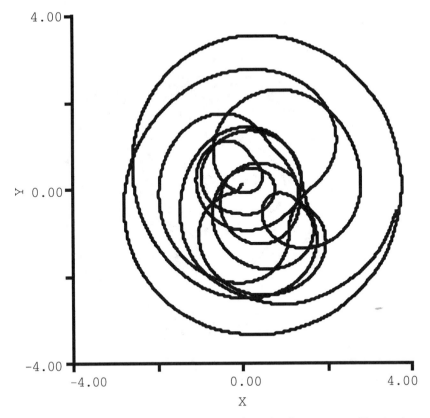

FIGURE 5.15   The trajectory of a passive particle in the flow generated by the three-vortex system.

with bands that are imagined to extend infinitely far to the north and south. Within each band, the vorticity is uniform and is equal to the northward velocity on the east side of the band minus the northward velocity on the west side divided by the width of the band. Equivalently, it can be said that the vorticity within the band induces a change in the northward velocity across it. A band with positive vorticity then causes the bands to its east to slide northward and causes those to its west to slide southward.

The choice that the bands are oriented south to north is not arbitrary. On a rotating planet, the component of planetary rotation that is oriented along the local vertical changes with latitude—the so-called beta-effect, discussed in Section 4.5. When it is at rest relative to the planet, air close to the North Pole has greater vorticity (referring strictly to the vertical component) than air close to the equator. When air flows poleward, its deficit of vorticity appears as a tangible negative (clockwise) rotation, and vice versa for equatorward

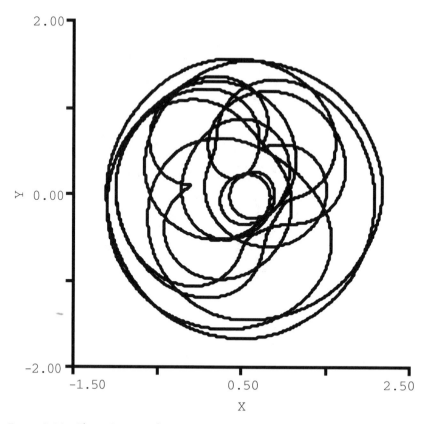

FIGURE 5.16   The trajectory of one vortex in a system with four equal vortices.

flow. Thus, if one band causes other bands to slide northward or southward, these bands then take on tangible vorticity and will cause other bands to move. As all the bands slide, they induce motion in all the other bands.

A model with 10 such bands is shown in Figure 5.18. The vorticity in each band increases, through the beta-effect according to whether the meridional velocity, v, is southward or northward. The value of v for each band depends on the vorticity of every band in the system. The system is set up so as to be relevant to motions along a constant latitude and extending around the globe. The constraint on the meridional velocities is that they must sum to zero; the vorticity of the bands cannot induce a uniform drift northward or southward. This constraint is enforced by the complicated expression for v1 in the converter near the bottom of the model. Once one value of v is given, the others are obtained by moving left to right through the model, adding the vorticity in each band, multiplied by its width. This is done in the row of converters along the bottom of the model.

The vorticity within each band can also be changed by a west-to-east flow through the band. This is an essential difference between this model

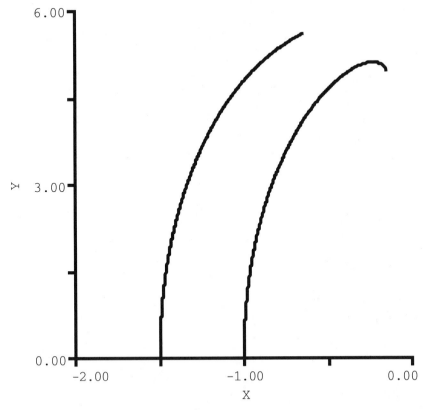

FIGURE 5.17   Trajectories of a pair of vortices of equal strength and with opposite signs, interacting with a vertical wall at x = 0.

and that for point vortices. The point vortices were "glued" to a bit of fluid and moved with it. Here, fluid can flow through the bands, and the model then treats the vorticity within the fixed region of the band as a stock. The former treatment of fluid flows is denoted *Lagrangian,* and the latter, *Eulerian.* The Lagrangian view, because of its similarity to the dynamics of solid objects, is more intuitive, but for complex fluid problems, the Eulerian approach is almost invariably taken, especially when the flow is to be modeled on a computer. The spatially periodic nature of this system, which makes it suitable for the behavior along a parallel of latitude on a spherical earth, is evidenced by the flow that advects vorticity between the first and tenth bands.

To see that this model supports waves, it is initialized with a sinusoidal distribution of vorticity with respect to *x,* the west-to-east coordinate. For convenience, the vorticity is converted into a meridional displacement, by dividing by beta. The values y_1 to y_10, represent the north–south distance the bars must slide to take on that value of vorticity. STELLA's graphing

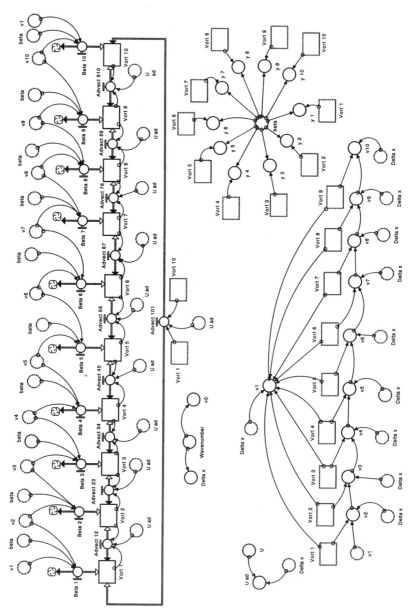

FIGURE 5.18   A model of the Rossby wave.

package is not well suited to representing the spatial behavior of variables, so it is best to save the model results in a table, for plotting using a graphing program. In Figure 5.19, the meridional displacements are shown as a function of longitude at three times, each separated by 1/2 day. The widths of the bands are chosen such that three of these models, and thus three wavelengths of the wave, would fit around a parallel of latitude. For simplicity, the zonal wind, U, blowing through the model from west to east, is set to zero. The striking result from this figure is that the wave moves to the *west* with time, in contradiction to common knowledge that weather patterns in middle latitudes typically move eastward. This discrepancy has two sources. First, this is a rather long wave, with a wavelength greater than the east–west extent of a typical weather system. Second, in contrast to the prevailing westerly winds of middle latitudes, here the zonal wind has been set to zero. Converting longitude at latitude 45°N to distance, it can be calculated that this wave moves westward at about 35 m/s. If this wave were imbedded in a strong westerly jet stream, say of 50 m/s, it could be carried to the east, as is shown in Figure 5.20.

Waves of this type are denoted *Rossby waves,* after Carl Gustav Rossby, the 20th-century Norwegian meteorologist who first explained their dynamics. How do the speed and frequency of Rossby waves vary with their wavelength? Most familiar waves, such as sound and light waves, are nondispersive, in that all wavelengths travel at the same speed, or nearly so. To explore this property for Rossby waves, a sensitivity experiment is performed varying the parameter Wavenumber. Wavenumber is the number of complete wavelengths that fit around a parallel of latitude. This is varied in the model by altering the width of each band. Figure 5.21 shows how the north–south displacement of a single band varies with time for wavenumbers 1, 2, and 4, with U again set to zero. The frequency of wavenumber 1 is twice that of wavenumber 2, which is again twice that of wavenumber 4. Thus, in the absence of an advecting zonal wind, the frequencies of Rossby waves vary inversely with their wavenumbers, or in proportion to their wavelengths. Because the speed of the wave is equal to its frequency multiplied by its wavelength, this implies that the westward speed of Rossby waves increases with the square of the wavelength, or, equivalently, it varies inversely with the square of the wavenumber. In the middle-latitude atmosphere with its prevailing westerly winds, this means that very long wavelength disturbances move to the west, disturbances of intermediate wavelengths remain nearly in place, and short-wavelength disturbances are carried eastward by the westerly winds. This behavior is evident in any sequence of upper air charts—weather maps for the midtroposphere. Small-scale disturbances can be seen moving eastward through a nearly stationary pattern of longer waves. If one takes an average over a period of, say, 10 days, the moving disturbances are averaged out, and the remaining pattern has a wavelength close to that at which the Rossby-wave speed equals zero.

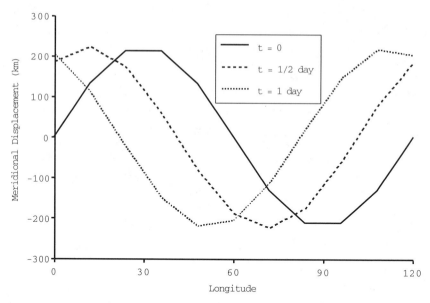

FIGURE 5.19   A Rossby wave, represented by meridional displacements, at three different times.

Dispersive waves, like Rossby waves, have the property that the speed and direction of energy propagation is different from the visible direction of motion of a sinusoidal wave, or, in other words, that the group velocity differs from the phase velocity. Figure 5.22 shows the result when, rather than starting with a sinusoidal disturbance, we start with a localized disturbance, neighboring bands with positive and negative vorticity. For simplicity, U is again set to zero. The southward displacement of the band at 60°E longitude grows, while the northward displacement of the band to its west decreases, and the bands further east move north. The energy of the disturbance is clearly propagating to the east, opposite to the direction that the peak or trough of a sinusoidal wave would travel. This may be understood by considering just the interaction between the two initially displaced bands. The negative vorticity in the western band creates a southward flow over the eastern band, increasing its southward displacement, but the positive vorticity of the eastern band creates a southward flow over the western band that decreases its northward displacement. Thus, the disturbed region moves to the east. This appearance of a new disturbance to the east of an existing one is a simple example of a phenomenon in the atmosphere known as *downstream development,* wherein a new storm system develops, as if by magic, to the east of an existing one.

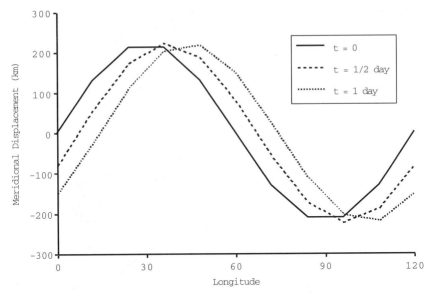

FIGURE 5.20   The same Rossby wave as in Figure 5.19, but now in a westerly flow of 50 m/s.

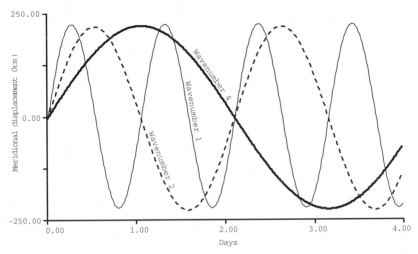

FIGURE 5.21   Meridional displacements at a point for Rossby waves with wave-numbers 1, 2, and 4.

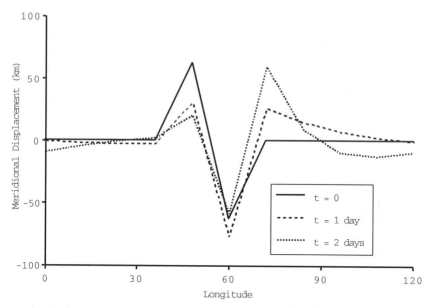

FIGURE 5.22   Meridional displacements, starting from a localized disturbance.

5.6 The Gulf Stream

The surface currents of the North Atlantic Ocean feature a narrow, intense, northward flow close to the American coast—the Gulf Stream—and a much slower southward return flow distributed over most of the remainder of the basin. The intensification of the strength of ocean currents along the western boundary is found further north in the Atlantic, where the southward-flowing Labrador Current hugs the Canadian coast, and also in the Pacific. It might be assumed that this westward intensification results from an east–west asymmetry in the winds that drive the currents. In fact, this phenomenon is a consequence of vorticity dynamics similar to those that produce Rossby waves, and despite the obvious two-dimensionality of the oceanic circulation, a sufficient model of the Gulf Stream (Figure 5.23) is very similar to the previous one for the Rossby wave.

Three changes have been made from the Rossby wave model. First, because coasts bound the ocean, there is no west-to-east flow advecting vorticity. Second, drag has been added that damps the vorticity in each band at a rate proportional to that vorticity. Finally, a forcing has been added to the vorticity in each band. This is denoted WSC, which stands for wind-stress curl, the tendency for the winds to make the underlying ocean currents at any point rotate in a counterclockwise sense. Because the winds over the subtropical Atlantic are dominated by the subtropical high-pressure system, they rotate clockwise and push the underlying ocean in the same direction. Thus, WSC is given a negative value. To demonstrate, however, that it is

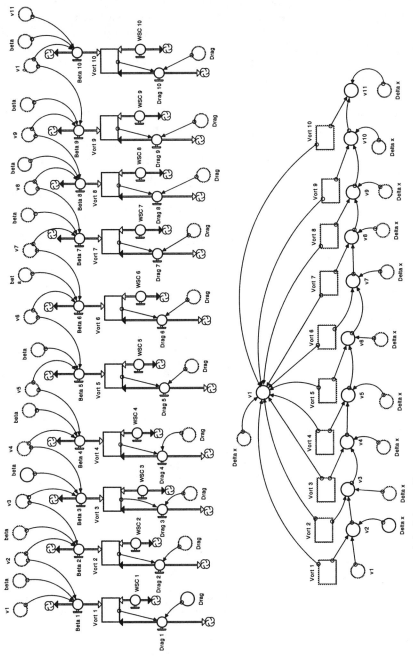

FIGURE 5.23   A model of the Gulf Stream.

not the distribution of forcing by the wind that causes the westward intensification, WSC is given the same value at every point. Note that only the
first 500 km of ocean east of the coast have been included in the model.
The bands must be kept sufficiently narrow to capture the Gulf Stream, so
the model must simply be made larger and include more bands, if the entire ocean is to be modeled. The choice of 10 bands here is due to the limited patience of the modeler rather than any technical restriction on the
construction of larger models.

The ocean starts at rest, and the uniform WSC is turned on at time 0.
Once again, to graph the results with respect to a spatial coordinate, it is
helpful to save the results in a table and plot them using a graphing application. Figure 5.24 shows how the north–south current is distributed across
the basin at different times in the simulation. Initially, a southward current
appears on the east side of the basin and a nearly symmetric northward
current appears on the west, consistent with the effort by the winds to
make the ocean rotate counterclockwise. By 20 days, however, the currents
are southward over much of the basin, and the northward current is
stronger and confined to the western third. Finally, in equilibrium, the westward intensification is even more pronounced. The northward current
along the western boundary is about one-tenth the width and ten times as
strong as the southward drift elsewhere.

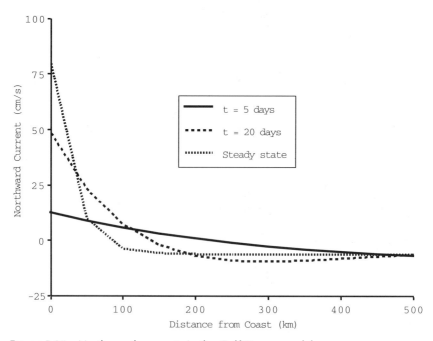

FIGURE 5.24   North–south currents in the Gulf Stream model.

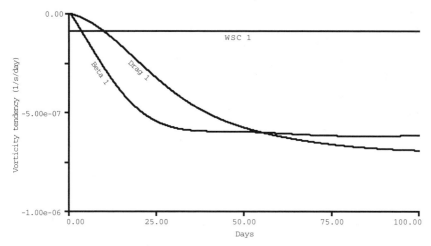

FIGURE 5.25   Terms in the vorticity budget for the westernmost band of the Gulf Stream model. (WSC, wind-stress curl.)

The dynamical balances are also entirely different within and outside of the western boundary current (the Gulf Stream). In the westernmost band (Figure 5.25), the northward advection of low vorticity, Beta_1, which tends to increase the counterclockwise rotation, nearly balances the frictional loss of vorticity, Drag_1, with a little help from the winds, WSC_1. Elsewhere (Figure 5.26), the vorticity is negligible, and so is the drag. A balance is, therefore, obtained between the effort of the wind to increase clockwise

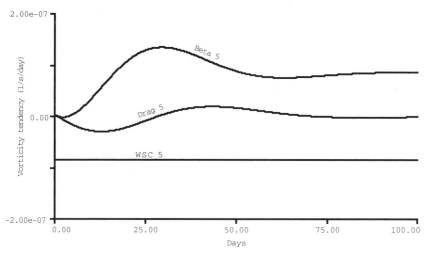

FIGURE 5.26   Terms in the vorticity budget for the fifth band of the Gulf Stream model.

rotation, WSC_5, and the advection of positive vorticity, the tendency toward counterclockwise rotation, from the north, Beta_5. This latter balance is termed *Sverdrup balance,* and it describes the mean currents over most of the ocean, away from the western boundaries.

The development of the asymmetry can be understood as follows. When the wind is "turned on," water starts to drift southward on the east side of the basin and northward on the west side. The water on the west, however, obtains negative (clockwise) vorticity both from the wind and from its northward motion. Its vorticity, which is manifested as strong clockwise shear in the northward current, increases until its two sources are balanced by drag. Elsewhere, however, the negative vorticity supplied by the wind is readily balanced by the southward drift, and drag need not enter the picture.

Problems

5.1 Starting with the second model of convection (Section 5.2), develop a model of the sea breeze. The sea breeze arises along an ocean or lakeshore (or even that of a large river) when the sun heats the shore. The water warms more slowly than the ground because of its greater heat capacity. The warmth of the land is conveyed to the overlying air by convection and mixing. Thus, the key element of the sea breeze is that it is a circulation driven by a lateral rather than a vertical contrast in the heating of a fluid.

To convert this model into one for the sea breeze, first, eliminate the horizontal and vertical conduction, which are not important for this circulation. Now, relabel the stocks. The heated stock, which was the Bottom, becomes the Land, the Right becomes the Top, the Top becomes the Sea, and the Left becomes the Bottom. The circulation is driven by the contrast between the Ocean and the Land. Therefore, it is necessary to change the prescription of Delta_T accordingly.

For physical dimensions, the imposed temperature contrast, Delta_T_0 should be on the order of a few degrees. The Land and the Ocean should be heated and cooled over a timescale of hours; the vertical extent of the circulation, H, should be about a kilometer; and the Drag should act on roughly the same time scale as the heating and cooling.

Run your modified model, and ascertain that the term "sea breeze" is appropriate. How does the presence of the sea breeze modify the temperature over the land?

5.2 The multiple-vortex model (Section 5.4) can provide almost endless numbers of esthetically pleasing and interesting results (see Aref,[1] for a comprehensive review of the possibilities). Some interesting things to try are the following:

a. A set of three vortices of equal strength and sign start out equally spaced along a line. In this configuration, they will rotate forever around the center vortex. Is this a stable configuration? In other words, does the relative arrangement of the vortices remain approximately the same if the central vortex is displaced slightly from being centered between the other two?

b. Repeat step a, but now give the center vortex a circulation of the opposite sign. How does this change the nature of the instability?

c. Set up a vortex pair so that it "collides" with a single vortex. What happens if the collision is "head on"? If the collision is glancing? There is a strong analogy here with the collisions of particles.

d. Set up a collision between two equal and opposite vortex pairs. What happens?

5.3 In the discussion of the multiple-vortex model (Section 5.4), a connection was drawn between the chaotic evolution of the four-vortex model and its unpredictability. *Unpredictability* means that the state of the system after some time is very sensitive to its initial state. Run a chaotic case of the four-vector model. A useful chaotic case can be constructed by using four vortices of the same sign and equal strength. Run the model, then slightly alter the initial position of one vortex and run the model again. Make a comparative plot to track one coordinate of one of the vortices in the two different cases. Do the two solutions track each other forever, or do they diverge? How does your result change when the experiment is repeated for a three-vortex model (i.e., when the strength of one vortex is set to zero)?

5.4 In the Rossby wave model (Section 5.5), explore what value of the mean wind, U, is required to "freeze" the Rossby wave in place. In this case, the flow undulates with longitude but does not change in time. Because it does not propagate away, this is the scale of wave that shows up when weather maps are averaged over many successive days, and this is the scale associated with persistent patterns of weather. How does the wind speed required to freeze the wave vary with the wavenumber? Typical mean winds in the winter lower atmosphere are on the order of 10 to 20 m/s but are much higher, 60 to 100 m/s, in the stratosphere. What do your results then suggest about the wavelengths of stationary patterns in the stratosphere compared with the troposphere?

5.5 In the Gulf Stream model (Section 5.6), a constant value of WSC was used across the ocean basin. Is this essential? Is a Gulf Stream, a concentration of northward flow along the western boundary, still obtained for different distributions of WSC? What if the only nonzero value of WSC is on the eastern edge of the basin?

Further Reading

The two books cited at the end of Chapter 4 provide, in the case of Holton's text, a good introduction to the circulation, and in Cushman-Roisin's book, physically based discussions of Rossby wave propagation and the Gulf Stream. Many different examples of convective flows are described in *Buoyancy Effects in Fluids,* by J. S. Turner (1973, Cambridge University Press, 368 pp.). This book features a wonderful collection of plates depicting many of these flows in the laboratory and in nature.

References

1. Aref, H., 1983: Integrable, chaotic, and turbulent vortex motion in two-dimensional flows. *Annual Reviews of Fluid Mechanics,* **15**, 345–389.

6

Dynamical Models of the Climate

6.1 The Hadley Circulation

In Chapter 2 we described models of the climate in which heat and moisture moved from one place to another diffusively. In the real climate system, these transports are dynamically controlled and are much more interesting. As in the simple diffusive climate models, more energy arrives from the sun in the tropics than is lost to space as infrared radiation, whereas in high latitudes, more energy is lost to space than is gained from the sun. Thus, it is necessary that the tropics export energy to higher latitudes. How energy is transported poleward depends on the latitude. The first step in the process, getting the energy from the tropics to the subtropics, is accomplished by an overturning circulation in the latitude–height plane called the *Hadley circulation*.

Before going on, a note on terminology: In meteorology, winds are labeled according to the direction from which they blow. Thus, winds that blow from the west toward the east, in the positive x direction in our usual coordinate system, are denoted *westerlies,* and winds that blow from the south to the north, in the positive y direction, are denoted *southerlies.*

We consider air that is heated in the tropics and rises. In the upper troposphere, it spreads poleward into both hemispheres, until it reaches the subtropics, where it sinks and flows back into the tropics near the ground. A schematic of one hemisphere of this circulation is shown in Figure 6.1, which also shows the three dynamical balances that keep it in equilibrium. The temperature in the tropics represents a balance between heating and the adiabatic cooling due to rising motion (see Sections 3.2 and 3.3), while the cooler temperature in the subtropics represents a balance between radiative cooling and warming by descent. The overturning circulation is driven by the tropical/subtropical thermal contrast, much like in our models of convection (Sections 4.1 and 4.2), but it is opposed by friction and by the Coriolis deflection of the zonal winds—the westerlies aloft and the easterlies at the surface. The westerlies aloft are enhanced by the advection of high angular momentum from lower latitudes, since on a rotating planet, air at rest relative to the planet has the greatest angular momentum at the

121

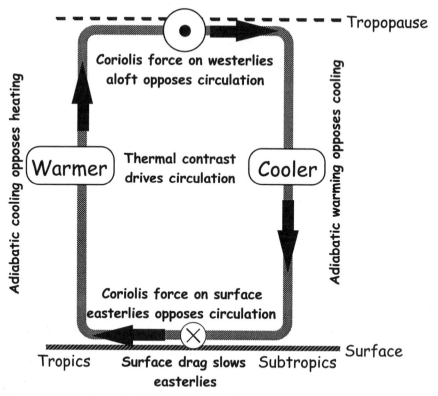

FIGURE 6.1   A schematic of the Hadley circulation.

equator. The surface easterlies are maintained against surface drag by the advection of lower angular momentum from higher latitudes.

These are the dynamics represented in this model (Figure 6.2). It is necessary for the model to keep track of the temperatures, the strength of the overturning circulation, and the angular momentum associated with the zonal winds. These three interacting systems are shown in the top half of the model. The converters and connections on the lower right are required to relate angular momentum, the conserved quantity represented in the stocks, to the observed quantity, the wind. The heating and cooling are treated as linear relaxations toward reference temperatures, and the vertical and meridional flows are such that mass is conserved as the air circulates around the cell. A question arises, however, as to whether all this is necessary. Consider a state in which there is no circulation around the cell and no wind at the surface and, therefore, no effect of surface friction. In this case, the temperatures would be equal to the reference temperatures (since there would be no rising or sinking) and the Coriolis force on the westerlies aloft would have to be sufficiently strong to exactly balance the tendency of the thermal contrast to drive the circulation. This balance is, incidentally, the geostrophic balance described in Chapter 4. Such a balanced

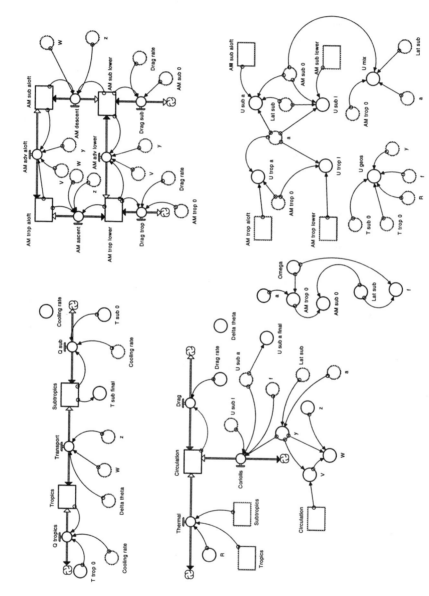

FIGURE 6.2 A model of the Hadley circulation.

state without a Hadley circulation is not observed in the atmosphere, but why not?

If we run this model from an initial condition in a state of balance as described above, it remains in that state forever. The temperatures remain in radiative-convective equilibrium, there is no wind at the surface or at the equator, and the wind aloft in the subtropics remains at its geostrophic value, the value for which the equatorward Coriolis deflection is exactly balanced by the poleward pressure gradient. For some settings of the parameters, if the initial condition is not in this perfect state of balance, the departure from balance increases until a new equilibrium is achieved. The evolution of the subtropical winds aloft, U_sub_a, is shown in Figure 6.3 for a case in which the Hadley cell is assumed to extend from the equator to 25°N, and the radiative equilibrium temperatures are 30°C at the equator, and 20°C at 25°N. The initial conditions are all in equilibrium, except that the initial value of the subtropical wind aloft, U_sub_a, is 1% below its equilibrium value. The system evolves, over a time of tens of days, to a new equilibrium, in which the subtropical wind aloft is substantially weaker. It approaches a value, denoted U_mix, equal to what would be achieved by air that starts out motionless at the equator and conserves its angular momentum as it moves to the poleward limit of the Hadley cell. Figure 6.4 shows that the final equilibrium state includes a circulation about the cell, represented by the southerly winds aloft, denoted V, with rising motion at the equator and descent in the subtropics. The advection of angular momentum around the cell combines with the surface drag to produce weak easterly winds at the equator, both aloft and at the surface, and weak west-

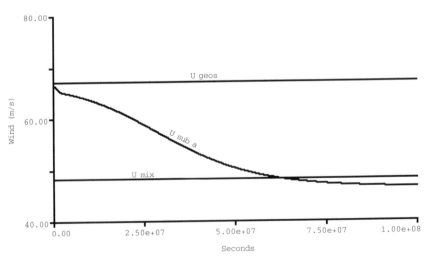

FIGURE 6.3   The upper-level subtropical zonal wind, U_sub_a in the Hadley circulation model, compared with the geostrophic wind in radiative equilibrium, U_geos, and that obtained from angular momentum conservation, U_mix.

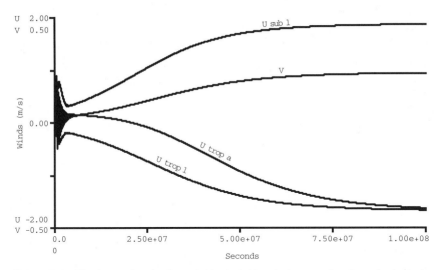

FIGURE 6.4   The lower-level subtropical wind, U_sub_l; upper-level tropical wind, U_trop_a; lower-level tropical wind, U_trop_l; and meridional wind, V.

erlies at the surface in the subtropics. Both are consistent with observed winds. Finally, Figure 6.5 shows that the cooling by rising motion in the tropics and heating by descent in the subtropics have reduced the temperature contrast across the cell from that in radiative equilibrium.

This circulation does not occur for all values of the parameters. For the present parameters, U_mix, which depends only on the latitude assumed

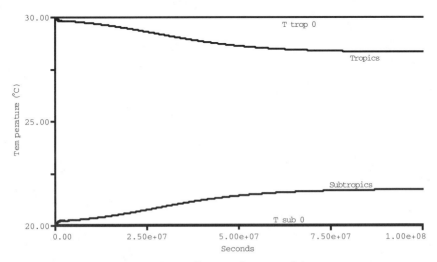

FIGURE 6.5   Temperatures in the Hadley circulation model.

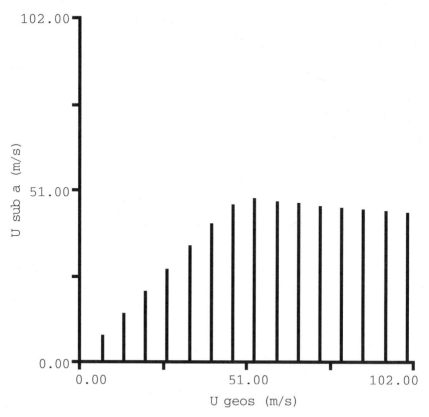

FIGURE 6.6   Subtropical upper-level winds, U_sub_a, versus their geostrophic value in radiative equilibrium, U_geos, for different values of the radiative equilibrium subtropical temperature, T_sub_0.

for the subtropical limit of the cell, has a value of about 48 m/s. A sensitivity run is performed, varying the radiative equilibrium temperature in the subtropics, T_sub_0, and running each case long enough that an equilibrium is reached. It is found that the final subtropical wind is equal to its geostrophic value when the geostrophic wind is less than U_mix (Figure 6.6). When the geostrophic wind is increased above this value—by reducing T_sub_0—a Hadley circulation appears and the subtropical wind no longer increases with decreasing T_sub_0. It decreases slightly with increasing geostrophic wind but remains in the neighborhood of U_mix. Figure 6.7 shows how the final subtropical temperature varies with its radiative equilibrium value. For values of T_sub_0 that are close to the radiative equilibrium temperature on the equator, there is no circulation, and the subtropical temperature is equal to T_sub_0; this is the right-hand portion of the graph where the slope is equal to 1. For lower values of T_sub_0, the Hadley circulation comes into play, warming the subtropics. Because the

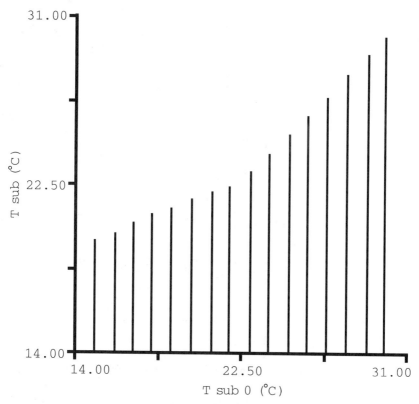

FIGURE 6.7   Subtropical temperature, T_sub, versus its radiative equilibrium value, T_sub_0.

circulation strengthens with increasing tropical–subtropical thermal contrast, the sensitivity of the subtropical temperature to T_sub_0 is reduced; this is the left-hand portion of the graph where the slope is approximately one-half.

In summary, a Hadley circulation is generated only when the radiative equilibrium geostrophic wind is greater than U_mix. It turns out that this is precisely the criterion that must be met for the radiative equilibrium geostrophic flow to be inertially unstable (Section 4.6): Parcels displaced poleward experience an increase in the poleward pressure-gradient force that exceeds the increase with latitude in the restoring Coriolis force. When the Hadley cell is established, and the subtropical wind is close to U_mix, the flow is nearly neutral to inertial instability.

Of course, in the atmosphere, there is not just one circuit of flow that defines the Hadley circulation. Rather, the circulation is distributed continuously over a large region of the tropics and subtropics. Two competing effects determine how far poleward the Hadley cell extends from the equator:

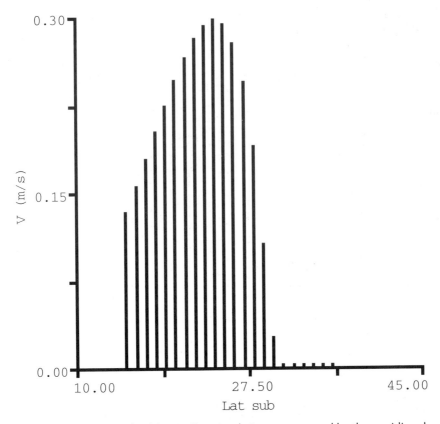

FIGURE 6.8   The strength of the Hadley circulation as measured by the meridional velocity, V, versus its subtropical limit, Lat_sub.

On the one hand, the radiative equilibrium temperature decreases with latitude, and this favors the development of the Hadley cell. On the other hand, the Coriolis force increases in strength with latitude, and this prevents the formation of a Hadley cell that extends all the way to the poles. Figure 6.8 shows the meridional velocity, V, a measure of the strength of the Hadley cell, in a sensitivity run in which the latitude of the poleward limit of the circulation, Lat_sub, is varied. The radiative equilibrium temperature in the subtropics, T_sub_0, is assumed to decrease with latitude in proportion to the sine of latitude squared. For the parameters used here, the Hadley cell can extend only to 32°; for higher values of Lat_sub, V = 0, and there is no Hadley cell. The condition for the flow with no Hadley circulation to be inertially stable is that the geostrophic wind in radiative equilibrium, U_geos, exceeds U_mix, the zonal velocity of air brought from the equator to the subtropical limit of the cell. Because the radiative equilibrium in the subtropics decreases and the Coriolis parameter increases, U_geos changes little with latitude. U_mix, however, varies with the reciprocal of the cosine of latitude, and so rapidly overtakes U_geos as the latitudinal extent of the cell is increased, permitting an inertially stable geostrophic flow.

6.2 The Midlatitude Cyclone

The poleward extent of the Hadley circulation is limited by the earth's rotation. Yet temperatures at middle and polar latitudes are far higher than would obtain in radiative equilibrium. Somehow the atmosphere (and the ocean) transports heat poleward out of the subtropics beyond the poleward reach of the Hadley circulation. This is accomplished by the alternating high- and low-pressure systems familiar to anyone who lives outside of the tropics. That these systems do indeed transport heat poleward is evident even to a casual observer of the day-to-day weather, especially in wintertime. Before a cold front arrives (in the Northern Hemisphere), relatively warm winds blow from the south or southeast, while after the front passes, cold winds blow from the northwest. Thus, as it passes, the weather system exchanges warm air from the south with the colder air to the north, and in so doing transports heat poleward.

Discussion of the detailed fluid dynamics that determine the structures of these highs and lows, denoted *baroclinic* * *cyclones* (lows) and *anticyclones* (highs), is beyond the scope of this book and the modeling capabilities of STELLA. A basic understanding can be obtained, however, by recognizing that baroclinic cyclones are a form of convection. The key difference from the convection considered earlier (Chapter 5) is that for baroclinic cyclones the atmosphere is stably stratified in the vertical; that is, the potential temperature (Section 3.2) increases with height. Because the potential temperature decreases to the north, convection is still possible, but now the fluid circuits that accomplish this convection lie in a plane that rises gently to the north. For conceptual simplicity, but without losing any essential points, it helps to neglect the compressibility of the atmosphere and to consider an incompressible fluid in which the temperature increases rapidly with height and decreases much less rapidly with latitude. Figure 6.9 depicts such an arrangement. Consider fluid flowing in a circuit, such as that shown in the schematic, which slopes upward to the north but less rapidly than do the isotherms. The northward-flowing air rises gently, but it still crosses isotherms from warm to cool, so it is warmer and therefore less dense than the air around it. The opposite obtains for the gently sinking air on the southward-flowing side of the circuit. The circuit, therefore, has less dense air rising and denser air sinking, and the motion can be sustained by the gravitational torque about the circuit.

---

*A *baroclinic* fluid is one in which the density varies on surfaces of constant pressure. In an ideal gas, such as the atmosphere, this implies that the temperature varies on pressure surfaces.

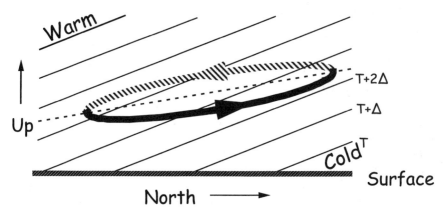

FIGURE 6.9   A schematic of the midlatitude cyclone as slanting convection.

The STELLA model of this system (Figure 6.10) closely resembles the model of convection in the previous chapter, only now, the reference temperature for the cold reservoir is determined by the north-to-south extent of the circuit, Delta_y; the slope of the circuit, Slope; and two parameters that describe the meridional and vertical distribution of the temperature before the onset of convection, DT_Dy and DT_Dz. With reference to the distribution of potential temperature in the earth's atmosphere, DT_Dz is chosen to be 3 K/km and DT_Dy is chosen as a 3 K decrease for every 1000 km northward displacement. When the model is started with a small initial circulation, the circulation may, depending on the value of the slope, amplify. If the amplification is assumed to be exponential, the growth rate, as in Section 5.2, is the log of the velocity after some time divided by its initial value. If the model is run for 1 day (86,400 seconds), this growth rate is in units per day. A sensitivity experiment can be performed to see how the growth rate depends on the slope of the circuit. The results are shown in Figure 6.11. The fastest growth, a rate approaching 1 per day (the amplitude increases by a factor of 2.718 in 1 day) is obtained for a slope half that of the isotherms. This is the optimal compromise between slopes that are too shallow, and for which the buoyant forces are nearly at right angles to the circuit, and slopes that are too steep. For excessively steep slopes, the circuit barely intersects the isotherms, and there is no buoyancy. Note that because of the very shallow slope of the isotherms, the slope for the most rapidly growing cyclone is necessarily also very shallow. Air rises only 0.5 km while traversing from the 2000 km from the south to the north side of the storm.

An interesting property of this model, which includes friction and the recharge of warm and cold air at the north and south sides of the storm and thermal damping at the east and west sides, is that it is mathematically equivalent to the famous Lorenz system of equations.[1] It was with this set of equations that Edward Lorenz discovered the phenomenon of deterministic

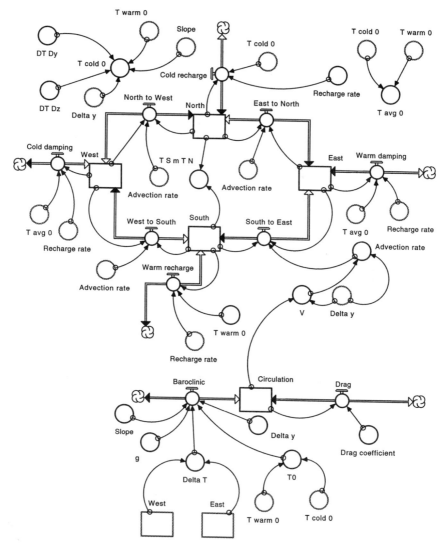

FIGURE 6.10   A model of the midlatitude cyclone.

chaos, the fact that the subsequent behavior of a simple dynamical system with a small number of degrees of freedom (this model can be constructed so that there are only three, rather than five, stocks) can exhibit extreme sensitivity to its initial conditions. The presence of chaos is associated with irregularity in the evolution of the system over time. Figure 6.12 shows the evolution of the velocity of the circulation, V, over more than 200 days. While there is a tendency for V to oscillate with a period around 8 days, it also makes irregular jumps from negative to positive values and back again. The classical depiction of the solution, the so-called Lorenz butterfly, is

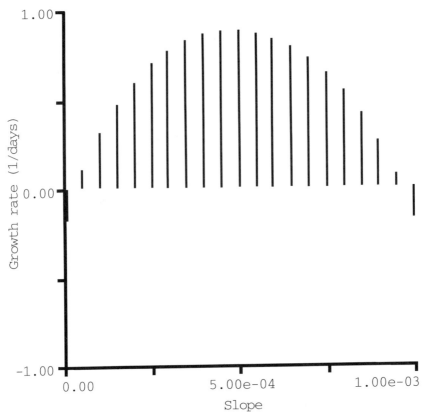

FIGURE 6.11   Growth rates of the midlatitude cyclone versus the slope of the circuit.

shown in Figure 6.13. Here, the temperature difference between the north and south is plotted against V. If one watches this graph as it is developing, one sees the trajectory orbit repeatedly around either the upper or lower "wings" of the butterfly, with occasional and irregular transitions between the two.

Recalling that the Lorenz system, in this case, is advertised as a model of a baroclinic cyclone, we can ask whether it does what baroclinic cyclones are supposed to do, namely, transport heat poleward. While the temperature varies irregularly, the model may still have a well-defined climate, defined by the time averages of the temperatures. As shown in Figure 6.14, the time-averaged north and south temperatures in this model approach nearly stable values after about 100 days. At these values, the north–south temperature difference has been greatly reduced, from more than 2.5 K to less than 0.5 K. Thus, the modeled cyclonic storm is, indeed, effectively transporting heat poleward.

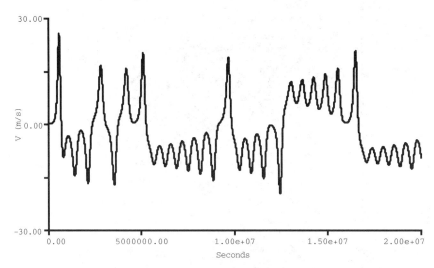

FIGURE 6.12   The meridional velocity in the midlatitude cyclone model.

It is well known that our ability to forecast the weather is limited. National and international forecasting centers issue forecasts of daily weather, based on the output from their numerical models, for no more than 10 days in advance. While these models can, of course, always be improved, it appears that there are fundamental limits on the range of useful weather forecasts. If the atmosphere is a chaotic dynamical system, and all evidence suggests that it is, then predictions from even a perfect forecasting model are limited by inevitable errors in the representation of the atmosphere that is given to the model as its initial condition. This effect can be explored in the present cyclone model by running the model repeatedly with slightly perturbed initial conditions. One such run is taken to be the actual evolution of the system, and the others are considered forecasts of that system, using a model that is, by definition, perfect. Figure 6.15 shows the results of forecasts of V in such a "perfect model" forecasting experiment. The initial conditions are taken from the middle of a long run of the model. The value of one variable, North in this case, is randomly perturbed in the initial condition of each run by an amount whose mean is zero and whose standard deviation is 0.1 K; this is easily accomplished using the "Sensi Specs . . . " and setting the "Variation Type:" to "Distribution." The solutions start to diverge significantly in this case after a million seconds, or about 12 days. By the middle of the simulations, or about a month, the different solutions appear to be nearly unrelated. At this time, in other words, a prediction from the forecast model is no better than a randomly selected state of the same system.

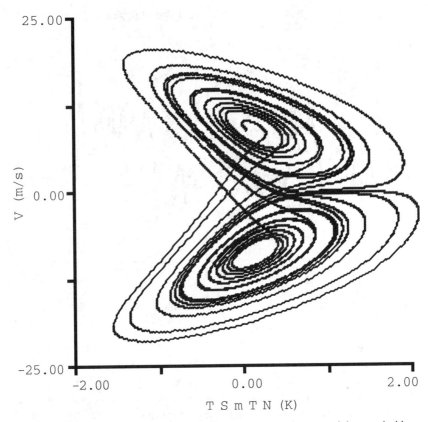

FIGURE 6.13 The Lorenz butterfly in the midlatitude cyclone model, revealed by plotting the meridional velocity, V, against the north–south temperature difference, T_S_m_T_N.

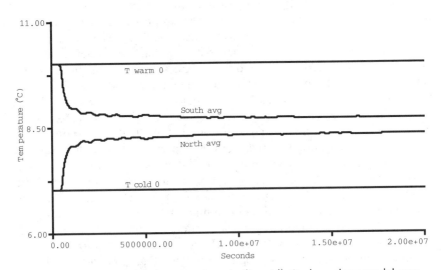

FIGURE 6.14 Time-averaged temperatures in the midlatitude cyclone model compared with their reference values.

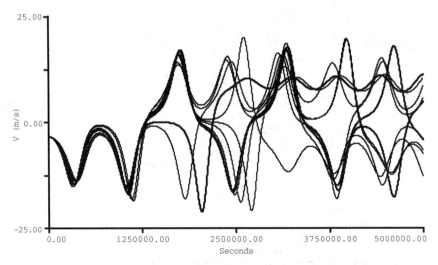

FIGURE 6.15   A set of "forecasts" for the meridional velocity in the midlatitude cyclone model.

6.3 Lorenz's General Circulation Model

Atmospheric general circulation models (GCMs) are the dominant tools in modern climate research. These models represent atmospheric winds, temperature, pressure, and humidity on a dozen or more atmospheric levels and at thousands of latitude–longitude locations. Newton's laws of motion for the atmosphere are solved at these "grid points," along with conservation equations for thermodynamic energy, atmospheric mass, and concentration of water vapor. The models include sophisticated treatments of the transfer of radiative energy through the atmosphere, of clouds, and of exchanges of heat, momentum, and moisture with the surfaces of the earth and ocean. While such models are impressive in their ability to simulate the most salient features of the earth's climate in significant detail, they are hugely demanding of computer resources; the models with the highest resolution (largest number of grid points) can be run only on supercomputers. There are only about 20 such models at research centers around the world.

The present model (Figure 6.16), due to Edward Lorenz,[2] is as simple as a GCM is complex, yet it includes some of the fundamental dynamical processes in the climate system. Much can be learned from observing and analyzing its behavior. The model has three variables: The first, Zonal_flow, represents the strength of the midlatitude westerlies. In the atmosphere, the strength of such flow is proportional to the rate at which temperature decreases with increasing latitude, so this same variable can be thought of as representing the magnitude of the equator-to-pole temperature contrast. The remaining variables, Sine and Cosine, represent different phases, with

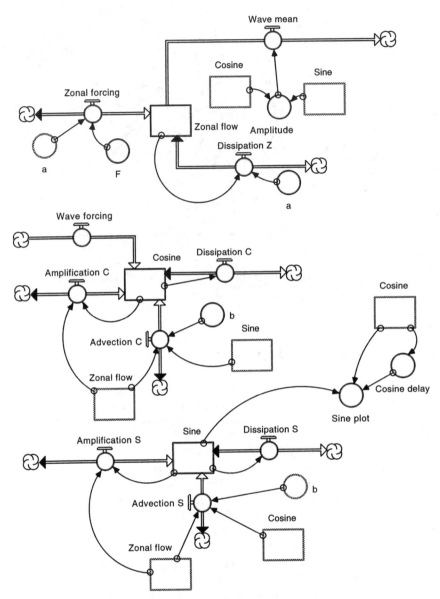

FIGURE 6.16   Lorenz's general circulation model (GCM).

respect to longitude of a single wavelike disturbance, a baroclinic wave. This wave is carried along by the zonal flow, so that there is a periodic exchange between the sine and cosine components at a frequency proportional to the strength of the zonal flow (the westward propagation of Rossby waves, Section 5.5, is neglected). The wave grows by extracting energy from the zonal flow, and it grows more rapidly when the zonal flow is stronger. At the same time, the wave weakens the zonal flow; its poleward transport of heat reduces the temperature difference between the equator and pole. Because the north–south winds in the wave and the size of its temperature perturbations are both proportional to its amplitude, the transport of heat by the wave varies with the square of its amplitude. Finally, the zonal flow is driven toward some equilibrium strength, a process that crudely depicts the forcing of the midlatitude westerlies by the heating and cooling of the atmosphere in high and low latitudes. Similarly, the wave is subject to some dissipation, representing the drag on atmospheric motions at the surface of the earth. With the dissipation rates set, as here, to unity, each unit of model time is approximately equal to a pentad—five days.

When the model is run in this configuration, two types of solutions are possible. When the equilibrium strength of the zonal flow, given by F, is below a certain value (equal to 1 for the parameters used here), the forcing of the zonal flow exactly balances its dissipation, and the amplitude of the wave invariably decays to zero. This is denoted the *Hadley regime,* because it corresponds to a climate in which waves play no role in the poleward transport of heat, and all heat is then necessarily carried poleward by the Hadley circulation. (This model, in fact, has no representation of the Hadley cell, so this name is something of a misnomer.) For larger values of F, however, the flow is unstable and the wave grows to some finite amplitude at which it can transport significant amounts of heat poleward. This leads to a strongly nonlinear response of the zonal flow to changes in its forcing, as is shown in Figure 6.17. Here we see the results of a sensitivity experiment in which F is increased from 0.2 to 2.0 in steps of 0.2. Values for each case are plotted after the model has reached steady state. In the Hadley regime, F < 1, the strength of the zonal flow is proportional to its forcing. Once F exceeds the critical value, the zonal flow remains at a value of 1 and is absolutely insensitive to F. Increases in the amplitude of the wave, and, therefore, its transport of heat, compensate for any further increase in forcing, as is shown in Figure 6.18. The amplitude of the wave is defined as the square root of the sum of the squares of the sine and cosine components. For values of F greater than 1, the Hadley regime zonal flow, which is prescribed as the initial condition (Zonal_flow = F), is unstable to the waves, and they rapidly grow to an amplitude that increases with F. Once equilibrium is reached, the wave propagates regularly, and the sine and cosine components vary sinusoidally in quadrature with each other.

In the atmosphere, especially in the Northern Hemisphere, heat is carried poleward not only by migratory cyclones and anticyclones but also by stationary disturbances that are generated by topography and by the thermal contrasts between continents and oceans. In Lorenz's GCM, this property is simulated by including some external forcing for one component of the wave. Interestingly, this effects a significant change in the behavior of the model. With a wave forcing equal to 1, an increase in F does not invariably lead to an increase in the strength of the zonal flow. Figure 6.19 shows the

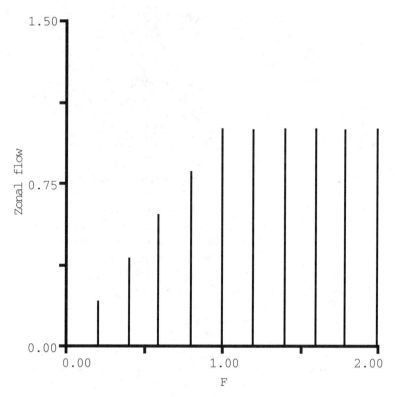

FIGURE 6.17   The zonal flow versus its forcing, F, in Lorenz's GCM, with
Wave_forcing = 0.

strength of the zonal flow at the end of 50 pentads in a sensitivity experiment in which F is increased in small increments. At values of F close to
1.5, the zonal flow becomes much stronger; it is readily ascertained that the
wave amplitude is small in these cases. The implication of this result is that,
because the atmosphere is a nonlinear dynamical system, it need not respond simply or smoothly to changes in its external forcing. In this simple
model, at least, an increase in the equator-to-pole temperature contrast in
radiative equilibrium can result in a *decrease* in the equator-to-pole temperature contrast realized by the climate.

The cases shown in Figure 6.19 have approached a steady state by 50
pentads. When the value of F is increased, however, unsteady solutions result, some periodic and some chaotic. Figure 6.20 shows a sample of such
chaotic behavior in the zonal flow and the wave amplitude, with initial
conditions taken from the middle of a longer integration. Here F = 5 and
Wave_forcing = 1. Bursts of wave activity rapidly weaken the zonal flow,
followed by periods when the waves are weak and the zonal flow gradually recovers. The zonal flow then yields to another burst of wave activity.
It has been suggested repeatedly over the last half century that such "relax-

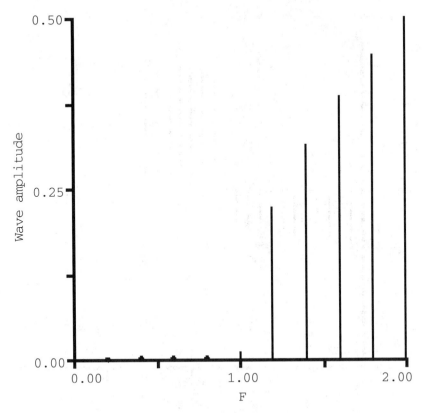

FIGURE 6.18   The wave amplitude versus the forcing of the zonal flow, F.

ation" oscillations, called *index cycles,* occur in the atmosphere. The idea is that pools of cold air accumulate in polar latitudes and then break out in a pulse of storm activity. This is an appealing idea, but observations have yet to confirm it unambiguously.

Even in a very simple system like this one with just three variables, chaotic solutions describe a complicated trajectory in a three-dimensional space. A standard approach to visualizing the structure of the chaotic solution is to construct a Poincaré section. To do so, a dot is drawn whenever the solution passes through a plane. In this case, we use the plane where Cosine = 0, and plot the values of Zonal_flow and Sine. In STELLA, this requires some effort. Using the DELAY function, the model compares the value of Cosine at the preceding timestep with its value at the current time. A new converter, Sine_plot is defined that is equal to the present value of Sine only when the preceding value of Cosine was less than 0 and the current value is greater than 0. Otherwise, Sine_plot is assigned a value off of the graph. Figure 6.21 shows the resulting Poincaré section for our chaotic solution. The trajectory of the solution rises through the plane in the handlike distribution of points on the right, and descends on the left. This complex

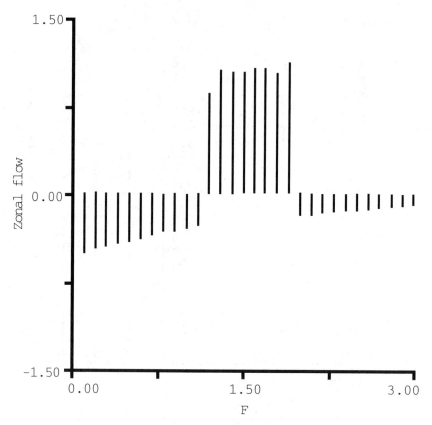

FIGURE 6.19   The zonal flow versus its forcing, F, with Wave_forcing = 1.

structure can be compared with a similar section for a periodic solution of the model, obtained when the Wave_forcing is reduced to 0.5 (Figure 6.22). Here, the trajectory of the solution intersects the plane only in the neighborhood of six points.

A final interesting wrinkle in this model is obtained by adding an annual cycle.[3] The forcing of the zonal flow represents the radiative driving of the equator-to-pole temperature contrast. This undergoes a strong seasonal cycle in nature, being stronger in winter than in summer. Here, F is made to vary sinusoidally between a wintertime value of 8 and a summertime value of 6,

$$F = 7 + 1 \times COS(2 \times PI \times TIME/73).$$

Because the unit of time in this model is a pentad, a year is 73 time units. When the model is run through several such annual cycles, it is found that different years are very different, especially during the summers—the periods when F is smaller. Figure 6.23 shows the variations of the zonal flow over eight annual cycles, with the sinusoidal variation of F shown for com-

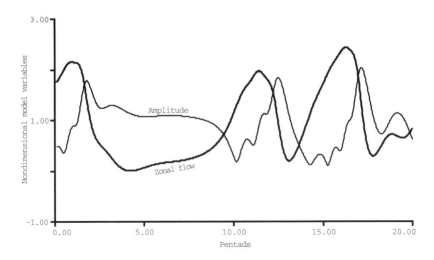

Fig 6.20

FIGURE 6.20   Chaotic behavior of the wave amplitude and zonal flow with F = 5, and Wave_forcing = 1.

parison. During some summers active variations in the zonal flow occur, while during other summers, the fourth, fifth, sixth, and seventh in this case, the variations are more rapid but weaker. The eighth summer shows a return to large amplitude variations.

What is the source of this remarkable behavior? As shown, the model is chaotic when F is at its wintertime value. For the summertime value of F, the solutions are periodic. There is not one such solution, however, but two. Figure 6.24 shows two solutions with F at its midsummer value of 6 but with different initial conditions. For both solutions at t = 0, Sine = 0 and Cosine = 1. The slow, strong oscillations are obtained when the initial value of Zonal_flow = 0, while the weak, rapid oscillations are obtained with an initial value of 1.2. These two solutions display the characteristics of the differing summers in the seasonal experiment. A system that realizes two different solutions, depending on the initial conditions, for a single set of parameters is called *intransitive*. This model is transitive in winter but intransitive in summer. Because the wintertime solution is chaotic, the model arrives at summer in a more or less randomly selected state. Thus the "initial condition" for the summer is determined by the winter's "roll of the dice." Sometimes it casts the system into the weak, rapidly oscillating solution and sometimes into the strong slowly oscillating solution. While this is an intriguing model for the interannual variability observed in the climate, it is not clear how it can be tested against observations. It is known that the real climate is periodic in no season, but it is possible for a system to be intransitive with two distinct chaotic solutions.

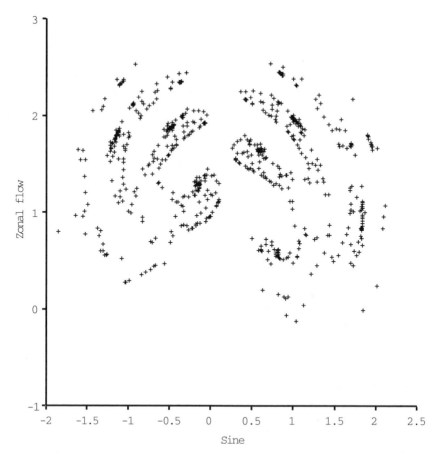

FIGURE 6.21   A Poincaré section for a chaotic solution of Lorenz's GCM with F = 5, and Wave_forcing = 1.

6.4 Chaotic Advection

In Lorenz's GCM, interesting behavior is associated with the chaotic dynamics of the dynamical system. Chaotic-looking behavior, however, need not necessarily be associated with chaotic dynamics. The concentrations of conserved or nearly conserved constituents of a fluid can take on beautiful turbulent-looking distributions, even though they are advected by flows that are simple and not at all chaotic. Figure 6.25 shows a satellite image* of

---

*The image in Figure 6.25 was obtained from the water-vapor channel of GOES-8 (Geostationary Operational Environment Satellite), operated by the National Oceanic and Atmospheric Administration. Such images can be obtained from the Web site of the Department of Atmospheric Sciences at the University of Illinois: http://ww2010.atmos.uiuc.edu/(Gh)/wx/satellite.rxml.

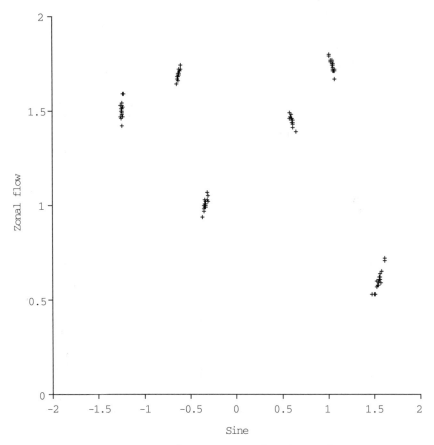

FIGURE 6.22   A Poincaré section for a quasi-periodic solution of Lorenz's GCM with F = 5 and Wave_forcing = 0.5.

the midtropospheric concentration of water vapor over North America and the Atlantic Ocean. Many swirls and eddies are visible, down to the smallest resolvable scales. Of course, the atmospheric winds that created this pattern are indeed chaotic, but in fact, even if they were not, the appearance of the water-vapor distribution might be quite similar. From a practical point of view, we care about the distribution of water vapor because it is the essential precursor of rain or snow. A midlatitude cyclone creates rain when it entrains a significant amount of water vapor into its rising motion. Figure 6.25 demonstrates why it is much harder to forecast the amount of rain than, say, the sea-level pressure. The amount of rain depends on the amount of water vapor in the air, and the amount of water vapor varies significantly on smaller scales than is resolved by the weather-balloon network or within computer forecasting models.

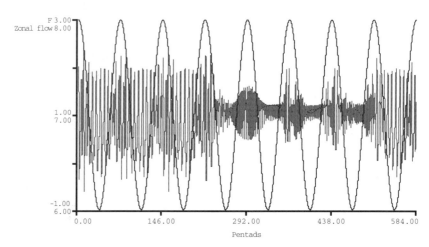

FIGURE 6.23   Results from Lorenz's GCM with an annual cycle in the forcing of the zonal flow, F. The sinusoidal curve shows F, and the irregular curve shows the strength of the zonal flow.

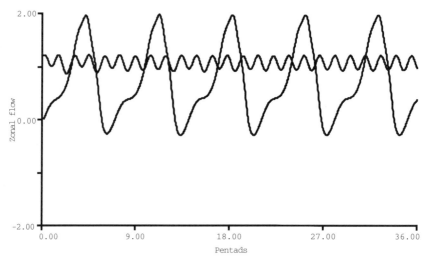

FIGURE 6.24   Summertime behavior of the zonal flow with two different initial conditions.

To understand this process, we consider a model (Figure 6.26) in which a simple flow leads to a complicated distribution of a conserved quantity. (This model is very similar to one described by Pierrehumbert.[4]) We consider a rectangular domain of earthlike dimensions, 30,000 km east to west, and 10,000 km north to south. For plotting purposes, the domain is taken as periodic in the east–west direction, so we can think of air blowing around

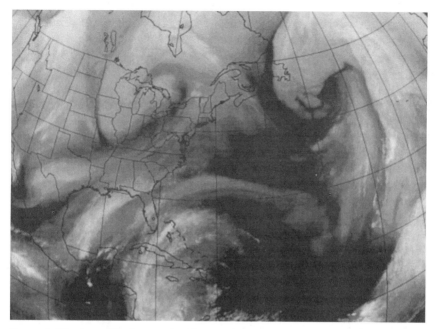

FIGURE 6.25   A satellite image of midtropospheric water vapor concentrations over the North Atlantic Ocean.

the globe, primarily from west to east. This has no effect on the actual calculations. The stock, x, represents the east–west position of a particle as though its trajectory were "unwound" from the periodic domain. The flow has two components. Its steady part has mean westerly winds blowing through a stationary wave. The streamlines for the steady flow are shown in Figure 6.27. These are the curves along which the fluid flows, so that the steady part of the velocity is everywhere parallel to these curves. Note that near the northern and southern boundaries of the domain there are regions of locally closed streamlines. Note also that this streamfunction does not appear explicitly in the model. In fact, these streamlines are plotted by following the positions of particles as they are carried along by the flow. The model needs only the west and south winds to advect particles. The winds are obtained by differentiating the streamfunction with respect to x and y. Obtaining the winds in this fashion assures that the motion is nondivergent and that advected particles do not pile up in one particular spot.

Superimposed on the steady flow is a traveling wave with the same wavelengths in x and y as the wave in the steady flow. Figure 6.28a and b show the streamfunction for the total flow, traveling plus stationary parts, when the wave is in opposite phases. With the parameters chosen, the traveling wave is stronger than the stationary component and dominates the combined flow. For the results shown here, the period of the wave

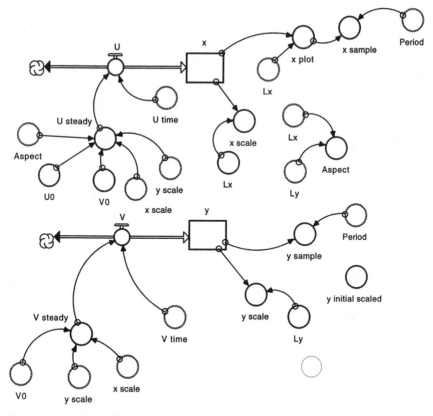

FIGURE 6.26   A model of chaotic advection.

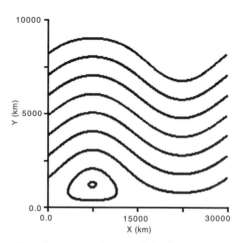

FIGURE 6.27   Streamlines for the steady part of the flow in the chaotic advection model.

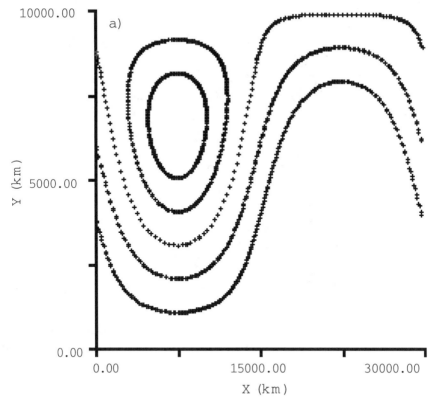

FIGURE 6.28   Streamlines for the time-varying flow at opposite phases. (b) shows a time 1/2 period of the traveling wave after that shown in (a). (*continued*)

is 5 days. While these details are important, the essential characteristic of this system is that the flow is simple in its spatial structure, it has no small-scale variations, and its variations in time are periodic, and thus regular and predictable.

The model executes the very simple task of following particles that are advected by the flow, essentially following labeled particles of air through the domain. In a steady flow, the particles follow the streamlines. When the flow varies with time, the trajectories of parcels become spirals that follow the advection by the steady part of the flow (Figure 6.29). What is not apparent from such trajectories is how the character of the advection varies dramatically as the starting position of the particle is changed. This can be shown by Poincaré sections of the positions, obtained by plotting the position at the beginning of each oscillation of the time-varying part of the flow.

Figure 6.30 shows such sections for three different initial positions along the line $x = 7500$ km, at $y = 1500$, 1875, and 2250 km. The model is run for 4005 days, and the position is plotted at the beginning of every fifth day.

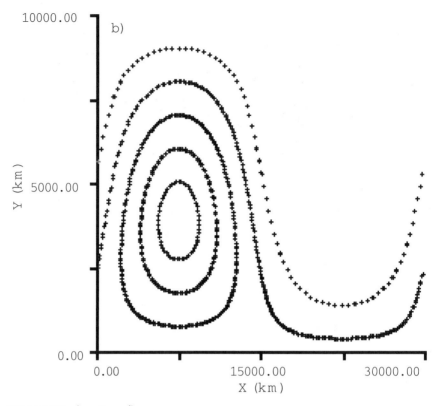

FIGURE 6.28   (*continued*)

For the first and third initial positions, the particle's position at the beginning of each period lies along a curve or a set of closed curves. For the
particle starting at $y = 1875$ km, however, the trajectory appears to visit
points that fill an area of the domain. The space-filling character of this trajectory can be confirmed by extending the run from this initial condition
beyond the length, 4005 days, permitted by STELLA. This is accomplished,
somewhat tediously, by displaying the end position at the end of the run as
a number, and then initializing the next 4005-day run with this position by
hand.

Figure 6.31 shows a run extended in this way to 24,030 days. Here, it is
clearly seen that the trajectory is indeed space filling, occupying a doilylike
region. Within the holes are "islands of stability," where the trajectory repeatedly visits only those points lying along a few closed curves. Bear in
mind that this complicated behavior of the advected particles results from a
fluid flow that has only periodic variations in time, and similarly simple and
large-scale variations in space.

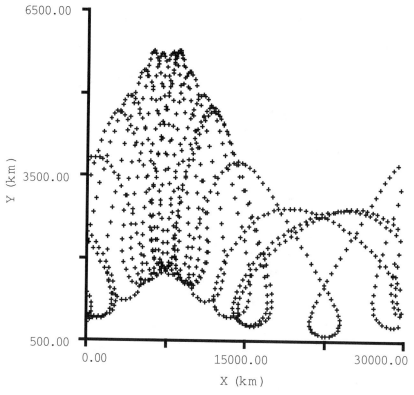

FIGURE 6.29   A parcel trajectory in the chaotic advection model.

The solution displayed in Figure 6.31 is an example of chaotic advection or, more generally, of Hamiltonian chaos. This is chaotic behavior exhibited by a conservative system, one lacking in any dissipation. Hamiltonian chaos appeared earlier in the multiple vortex model of Section 5.4, and it is distinct from the chaos exhibited by forced dissipative systems, such as the cyclone model and the Lorenz GCM in this chapter, and, presumably, the atmosphere.

While the patterns produced by chaotic advection are visually appealing, the practical implications are not immediately evident from Poincaré sections. Chaotic advection, however, shares the property with the chaotic dynamical systems described in both this and the preceding chapter that the solution is highly sensitive to the initial conditions. In this case, the implication is that the positions of initially nearby fluid particles will rapidly diverge. In this model, the positions of two particles can be followed simultaneously by copying the entire model and pasting it within the same model. Note that STELLA automatically relabels all the names by adding a

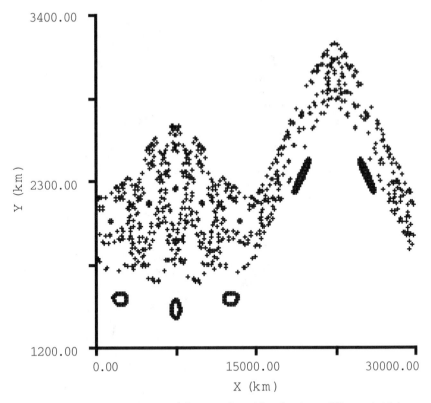

FIGURE 6.30   Poincaré sections of the parcel position for three different initial positions.

"2" to the end of each name. Now the separation between particle 1 and particle 2 can be calculated using the Pythagorean formula. The initial particles are given the same initial y positions, and are given x positions that differ by just 1 km. The experiment is carried out for two initial values of y: The first corresponds to the islands of stability shown in Figure 6.30, with y = 2225 km, while the second corresponds to the chaotic result shown in Figure 6.31, with y = 1875 km.

The results are shown in Figure 6.32. For the solution in the islands of stability, the increase in particle separation with time is imperceptible, and the separation curve hugs the bottom of the graph. In the chaotic case, the particles separate rapidly, though not monotonically. The implication is that a string of particles, a material line of the fluid connecting these two particles, would be rapidly stretched, bringing the fluid particles initially along this line into close contact with particles that were initially far away. This is the process of chaotic mixing, important for the redistribution of water vapor and other chemical tracers, such as ozone, in the atmosphere.

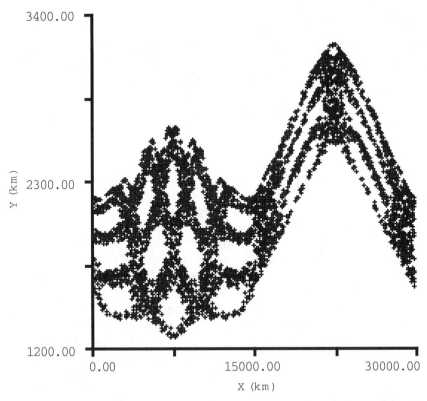

FIGURE 6.31   A Poincaré section for an extended run of the chaotic advection model.

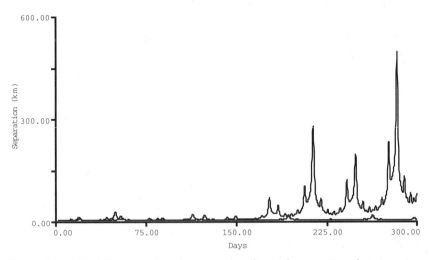

FIGURE 6.32   Particle separations for two pairs of particles, one in a chaotic region and the other in an island of stability.

Problems

6.1 Do you expect the Hadley cell to extend further poleward, or not as far, on a more slowly rotating planet? Check your reasoning by repeating the model runs used to generate Figure 6.8 (Section 6.1), using different values for Omega, the angular frequency of the earth's rotation. For reference, the earth has an angular frequency of $7.29 \ 10^{-5} \ s^{-1}$, while the values for Venus, Mars, and Jupiter are $2.98 \ 10^{-7}$, $7.05 \ 10^{-5}$, and $1.76 \ 10^{-4} \ s^{-1}$, respectively.

6.2 In the "forecasting" experiment for the midlatitude cyclone (Section 6.2), it was assumed that we have a perfect model for the system and that the initial error that subsequently grows to spoil our forecast is due only to uncertainties in initial conditions. Of course, our forecasting models are far from perfect models of the real atmosphere. Consider the same type of experiment as shown in Figure 6.15, only assume that we have perfect knowledge of the initial conditions but imperfect knowledge of the model parameters. As was done for Figure 6.15, repeat a model run with exactly the same initial conditions but with small perturbations to the value of some parameter, say the Recharge_rate. Does imperfect knowledge of the system lead to the same loss of predictability as does imperfect knowledge of initial conditions?

6.3 Construct a regime diagram for Lorenz's GCM (Section 6.3). Run the model repeatedly, varying the strength both of the zonal forcing, F, and of Wave_forcing. For each pair of values, note whether the solution is steady, periodic, or irregular. Draw a graph in which one axis is F, and the other is Wave_forcing. Indicate on the diagram the type of behavior obtained for each pair of values you have tried. What types of solutions occur at the extremes of your diagram—very large or small values of F and Wave_forcing? Can you understand them physically? How about in the middle of the diagram? Are the boundaries between different behaviors clear? Can the behavior of a run with a new pair of values generally be predicted in advance?

6.4 The association of chaos with the loss of predictability is a general property, not restricted to the midlatitude cyclone model described in Section 6.2. Perform similar predictability experiments as were used to generate Figure 6.15, but now using Lorenz's GCM (Section 6.3). Try using parameter values for which the solutions are periodic as well as irregular. How does the initial uncertainty grow in these different cases? Within the irregular cases, are some more unpredictable than others? Can you anticipate in advance, from examining a single run, which cases will prove less predictable?

6.5 Another way to visualize the chaotic advection (Section 6.4) is to follow the evolution of a set of particles initially lying in close proximity along a line segment. Using "Sensi Specs . . ." set the model to follow 100 particles starting at the same value of x but at values of y ranging over 10 km. Compare initial positions that are both within and outside of the range that yields chaotic trajectories. Modify the converter, x_sample, to look at the particle positions after 100, 200, and 300 days. (The positions of each particle at that time can plotted on a scatter plot with the "Comparative" feature selected.) How do the chaotic and nonchaotic cases compare? In the chaotic cases, in which direction, east–west or north–south, is the line of particles stretched more? Why? Particularly at 300 days, the final positions of the particles form clusters with intervening gaps. Can you describe what happens in the trajectory of two initially neighboring particles to give rise to such a gap?

To visualize what is happening, repeat the experiment, but with the time-dependent part of the flow turned off (V_prime = 0). In this case, where must a line of particles be initially located if it is to stretch rapidly?

Further Reading

The classic work on the dynamic circulation of the earth's atmosphere is the monograph by Edward N. Lorenz, titled *The Nature and Theory of the General Circulation of the Atmosphere* (1967, World Meteorological Organization, 161 pp.). Though out of print, this book is often available from libraries. It remains extremely useful and is probably the finest piece of scientific writing ever to emerge from the atmospheric or climate sciences. Other works by Lorenz are similarly valuable. His 1963 paper marks his discovery of chaos and its introduction into meteorology. He gives an informal account of that discovery in *The Essence of Chaos* (University of Washington Press, 227 pp.). Finally, his 1990 paper[3] provides an in-depth look at the properties of his GCM and a general discussion of irregularity and unpredictability in the atmosphere.

References

1. Lorenz, E. N., 1963: Deterministic nonperiodic flow. *Journal of the Atmospheric Sciences,* **20**, 130–141.
2. Lorenz, E. N., 1984: Irregularity: a fundamental property of the atmosphere. *Tellus,* **36A**, 98–110.
3. Lorenz, E. N., 1990: Can chaos and intransitivity lead to interannual variability? *Tellus,* **42A**, 378–389.
4. Pierrehumbert, R. T., 1991: Chaotic mixing of tracer and vorticity by modulated traveling Rossby waves. *Geophysical and Astrophysical Fluid Dynamics,* **58**, 285–319.

7

Climate Variations

7.1 El Niño

A fundamental property of the climate, whether at a specific locale, or globally averaged, is that it varies. The climate varies all the time and on all timescales. Some of these variations result from the internal nonlinear dynamics of the climate system, and some result from changes imposed externally, whether naturally or by humans, in the parameters that control the climate. This chapter presents models that represent some of the most important types of climate variability. The first is a model of El Niño.

El Niño is a coupled oscillation of the tropical Pacific Ocean and its overlying atmosphere. It is the single most important clearly distinguishable source of climate variability, not only on the shores of the tropical Pacific but over much of the world. Grain harvests in Zimbabwe are strongly influenced by El Niño, as is flooding in California.

The fundamental dynamics of El Niño are illustrated in Figure 7.1. The surface temperature of the tropical ocean is determined by the depth of the thermocline, the boundary between the warm waters near the surface and the cooler waters beneath. Along the equator, the slope of the thermocline across the Pacific is largely determined by the force of the winds blowing on the surface, called the *wind stress*. First consider the top panel in Figure 7.1, the state called La Niña, which is the opposite of El Niño. When the east-to-west wind stress associated with easterly trade winds is stronger than usual, water "piles up" in the western Pacific. The thermocline deepens, and there is a slight rise in the sea level. This rise in sea level is required if the pressure is to be horizontally uniform in the deep ocean. Since warm water is less dense than cold water, it takes a deeper column to make up a fixed amount of mass. Because temperature decreases with increasing depth in the ocean, a deeper thermocline implies warmer surface temperatures. Similarly, in the east, the shallower thermocline implies cooler surface temperatures. Because, all other things being equal, higher atmospheric pressures result from cooler temperatures, the atmospheric pressure is then higher in the east than the west. This makes the easterly trade winds blow harder at the surface, and piles up more

154

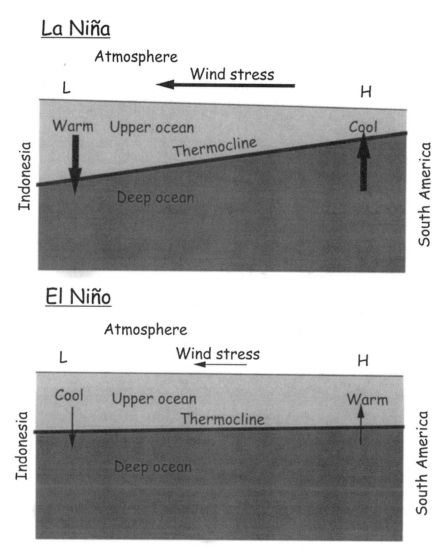

FIGURE 7.1   A schematic of El Niño and La Niña.

water in the west, deepening the thermocline, and so on. This is a positive feedback loop.

In El Niño, ocean temperatures are cooler than usual in the west and warmer than usual in the east. This results in a weakening of the trade winds, reducing the tendency for water to pile up in the west, making the thermocline shallower than usual in the west and deeper than usual in the east. These thermocline depths are associated with cool ocean temperatures in the west and warm temperatures in the east, so that the positive feedback loop is again closed.

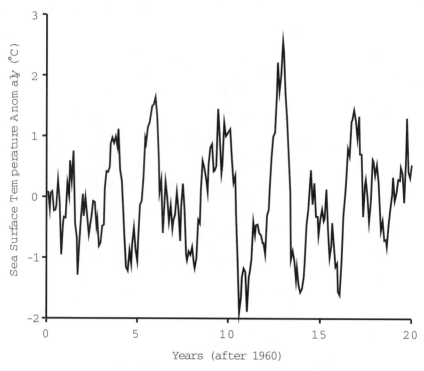

FIGURE 7.2   Twenty years of eastern tropical Pacific Ocean surface temperatures.

If this feedback loop were the whole story, we would expect the tropical Pacific to be locked into either an El Niño or a La Niña state, and only a very strong perturbation from outside the region would cause a switch from one to the other. This is not what is observed. On the contrary, El Niños occur irregularly but with an average frequency of about one every 3 to 4 years. For example, Figure 7.2 shows a 20-year record of the anomalous temperature (i.e., the deviation of the temperature from its long-term mean for that location and month of the year) of the ocean surface averaged over a large region of the eastern equatorial Pacific Ocean.* Warm temperatures in this graph correspond to El Niño events. There are perhaps seven distinct warm events in this record. Events vary greatly in their strength, their duration, and the time interval that separates them.

The model of El Niño described here and shown in Figure 7.3 was developed by Fei-Fei Jin.[1] In its unmodified form, it gives rise to periodic El

---

*These are data for the NINO3 region obtained from the Lamont Doherty Earth Observatory Web site:
http://ingrid.ldgo.columbia.edu/SOURCES/.KAPLAN/.Indices/.NINO3/.avGOSTA/

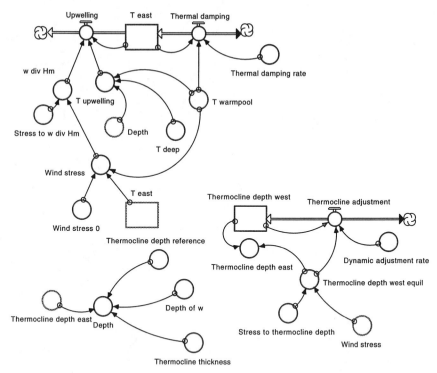

FIGURE 7.3 A model of El Niño.

Niños; we return to the issue of irregularity later. The positive feedback loop described above is contained in the part of the model that controls the evolution of East Pacific surface temperatures, T_east. Higher T_east gives rise to weaker east-to-west wind stress that reduces the cooling of T_east by upwelling. The strength of the upwelling is represented by the upwelling velocity scaled by the mean depth of the thermocline, denoted w_div_Hm. As the model runs through its El Niño–La Niña cycle, the wind stress varies in phase with T_east and out of phase with the upwelling velocity (Figure 7.4). Here, the times of higher T_east, less negative Wind_stress, and weaker upwelling (w_div_Hm), such as from day 500 to 1000, correspond to El Niño events.

By itself, however, the positive feedback shown in Figure 7.4 cannot explain the existence of self-sustaining oscillations. The key to the oscillatory behavior is in the second part of the model that describes the recharging of the deep layer of warm water (the so-called warm pool) in the west. When the westward wind stress is stronger along the equator (the La Niña phase), Sverdrup transport (see Section 5.6) causes water to converge on the equator in the upper ocean. This deepens the thermocline both in the east and the west and reduces the effectiveness of upwelling in cooling the east, since the cool subthermocline water is further from the surface. In the

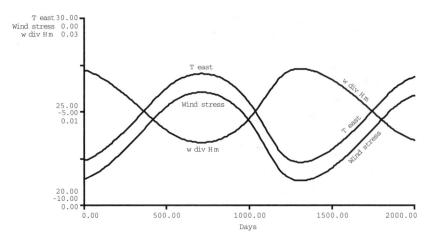

FIGURE 7.4   Eastern Pacific Ocean temperature (T_east), ocean upwelling (w_div_Hm), and eastward Wind_stress in the El Niño model.

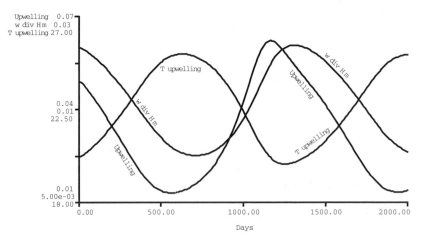

FIGURE 7.5   The temperature of upwelling water (T_upwelling), the strength of the upwelling (w_div_Hm), and the resulting cooling of the surface by upwelling (Upwelling).

model, Thermocline_depth_west influences Thermocline_depth_east linearly, and Thermocline_depth_east affects the temperature of the upwelling water, T_upwelling, through the converter, Depth. Figure 7.5 shows the strength of upwelling, w_div_Hm, the temperature of the upwelling water, T_upwelling, and the resulting cooling of T_east by upwelling, Upwelling, over one complete El Niño–La Niña cycle. As the strong upwelling (La Niña) part of the cycle proceeds from day 1300 to day 1500, the thermocline deepens raising the temperature of the upwelling water, T_upwelling, and thereby decreasing the effectiveness of upwelling in cooling the east, so Upwelling decreases. Similarly, in the weak upwelling, or El Niño, phase, the reduced recharge of thermocline water in the west eventually leads to a shallowing of the thermocline in the east. This cools the upwelling water, and makes the upwelling more effective in cooling the surface. The net effect is that the rate of upwelling lags behind the amount of cooling produced by this upwelling, and this time lag causes the model to oscillate. The lag is set by the time required for the thermocline in the west to recharge or discharge, denoted Dynamic_adjustment_rate in the model, so this parameter sets the duration of the El Niño cycles.

As noted above, observed El Niño cycles are highly irregular, quite unlike the periodic behavior of this model. What is the source of this irregularity? There are two very different paradigms for irregularity in climatic systems. One possibility is that the dynamics of the system are intrinsically irregular, that the behavior of the system is chaotic. The alternative is that the large-scale behavior of the system is regular, but that it is driven stochastically by the inherent irregularity of behavior on smaller scales. Before the discovery in the 1960s of deterministic chaos in systems with a few degrees of freedom (low-order chaos), it was believed that all irregularity in nature stemmed from the latter source. Subsequent to its discovery, however, many scientists have found low-order chaos an attractive explanation for observed irregularity and unpredictability, and the idea has probably been overapplied.

While some models of El Niño are in fact chaotic, especially when the annual cycle of background conditions is included, this particular model does not appear to support chaotic solutions. If the observed irregularity is to be simulated in this model, it must result from the application of stochastic forcing. The inclusion of stochastic forcing is not meant to imply that the climate system is other than chaotic. It is almost certain that atmospheric dynamics are chaotic, and this places fundamental limits on our ability to forecast the weather (see Section 6.2). Rather, the implication is that the dynamics on the scales explicitly represented in this model are not necessarily chaotic. The stochastic forcing can be viewed as a rough but convenient representation of the chaos known to be present on smaller scales or elsewhere in the system (e.g., in middle latitudes) within this model that represents only the largest scales of tropical motion.

A reasonable point at which to apply stochastic forcing in the El Niño model is in the wind stress. Physically, stochastic variations in the wind stress over the tropical Pacific could occur in association with outbreaks of cold air from continental Asia. These outbreaks are a feature of extratropical cyclonic storms (Section 6.2), so they could provide a way by which the chaos of midlatitude weather is communicated to the tropics. To add this stochastic forcing to this model, Wind_stress_0 is multiplied by a normally distributed random number at each time step,

Wind_stress_0 = −1 {°C} × (1. + NORMAL(0,Wind_stress_random)).

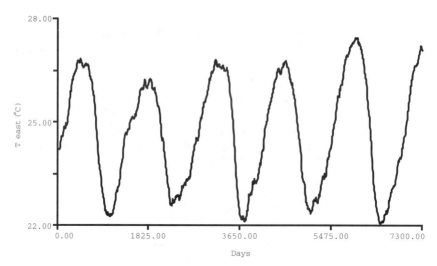

FIGURE 7.6   Eastern Pacific Ocean temperature, T_east, in the El Niño model with randomly varying wind stress.

When the model is rerun—with Wind_stress_random, the standard deviation of the stochastic variations, set to 0.4—the outcome is not encouraging. Figure 7.6 shows the results. The El Niño cycles are still very nearly periodic, though with some variations from one to the next.

Consider, however, the possibility that El Niño variations are not self-sustaining oscillations on the basin scale of the tropical Pacific but, rather, occur only because of rapid chaotic fluctuations in the atmosphere. To treat this case, it is first necessary to alter the model parameters so that it no longer generates self-sustaining oscillations. This may be done in many ways. Doubling the thermal damping rate, from 1/150 to 1/75 inverse days, is one. To maintain a reasonable temperature contrast between the East Pacific and the warm pool, it is then necessary to also double the baseline wind stress, Wind_stress_0, from 1 to 2°C. When this stabilized version of the model is run without stochastic forcing, its oscillations damp out in less than 20 years. When the stochastic forcing is applied (Wind_stress_random = 0.4), the result (Figure 7.7) is a much more realistic depiction of the temperature variations in the East Pacific. While a great deal of random variability occurs, the approximately 4-year periodicity of the El Niño cycle is still clearly evident. This model-generated record bears much similarity to the observed record shown in Figure 7.2.

The stochastically forced model also provides some insight into the predictability of El Niño and La Niña. If the behavior were perfectly periodic, it would be perfectly predictable. Although the El Niño cycle is the only part of the climate system that in current practice is skillfully forecast more than a season in advance, predictive skill largely vanishes for forecast times of a year or longer. To consider the predictability in our model, we make two

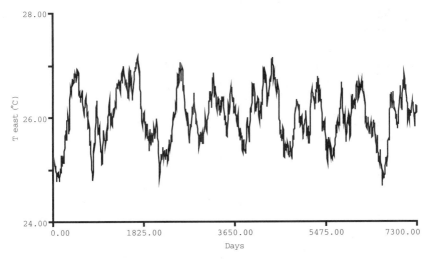

FIGURE 7.7   Eastern Pacific Ocean temperature, T_east, in the stabilized El Niño model with randomly varying wind stress.

assumptions. The first is that we have a "perfect model." We clearly do not have a perfect model of the actual El Niño, but the model is surely a perfect model of itself. In a predictability experiment, one run of the model is taken to be "reality," and additional runs of the model are considered "predictions." The second assumption is that the stochastic variations in the wind stress are completely unpredictable. This is a reasonable assumption. Atmospheric motions are predictable at most 2 weeks in advance, and we are interested in predicting El Niño a season or more ahead.

With these assumptions, all that is needed to run a predictability experiment is to run the model repeatedly from the same initial conditions but using different random numbers to generate the stochastic wind stress. Figure 7.8 shows the result of 15 such runs. The curves tend to follow each other for most of the first year, after which they disperse rapidly. When a sufficiently long time has passed, the model has "forgotten" its initial conditions, and any single run has no value as a forecast for another run. This can be quantified by calculating the standard deviation of the model prediction across a large ensemble of runs. (This is accomplished by saving the model from each run in a table, pasting the table into a spreadsheet, and using the spreadsheet to compute the standard deviation.) Figure 7.9 shows the standard deviation of 30 such runs as a function of the time into the forecast. Since all runs start from the same initial condition, the initial standard deviation is zero. The standard deviation grows rapidly during the first year of the experiment, after which, although it varies from month to month due to the relatively small size of the sample, it essentially saturates at a value of around 0.7°C. So, at least in this model, 1 year is about the outer limit for usefully skillful forecasts of El Niño.

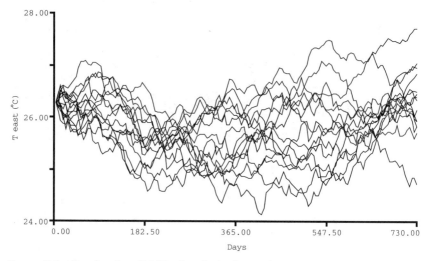

FIGURE 7.8   Results of an El Niño "prediction" experiment.

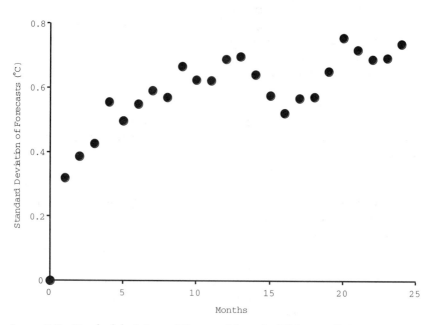

FIGURE 7.9   Standard deviations of "forecasts" from the El Niño prediction experiment.

It should be emphasized that the loss of predictability here is dynamically quite different from that displayed by the cyclone model in Section 6.2. There, the unpredictability arises from the chaotic nature of the model's behavior, combined with some unavoidable uncertainty regarding its initial state. Here, the model's behavior is completely regular. It is, in fact, boring. If it is not forced continuously, all variations damp out in few decades. The unpredictability comes from random forcing originating outside of that portion of the system that has been included in our model. Presumably, unpredictability of the latter kind can always be turned into the former by expanding the model to explicitly simulate the source of noise. Here, that would mean expanding the El Niño model to include a simulation of mid-latitude weather systems. Such an expansion is not always practical or desirable. A question for active research is whether random noise is really a good representation of the chaotic behavior in that part of the system we exclude.

7.2 The Quasi-Biennial Oscillation

When the El Niño model generates regular periodic oscillations, this is considered a defect. El Niño is not a periodic phenomenon. There are, in fact, almost no periodic phenomena in the climate system except for those that are driven by clearly periodic external forcing, such as the diurnal and seasonal cycles. A striking exception to the rule that variations in climate are aperiodic is found in the equatorial stratosphere, at altitudes between 18 and 35 km. Here, the winds undergo a regular cycle, as shown in Figure 7.10, with a period of about 26 months.[2] The figure shows 20 years of equatorial zonal winds in the stratosphere, averaged over all longitudes. The dark stripes in the figure represent westerly winds, and the light stripes are easterlies. The amplitude of these oscillations is about 20 m/s. Westerly winds develop in the middle stratosphere and then descend over time. Meanwhile, the westerlies aloft weaken and are replaced by easterlies, and these easterlies similarly descend with time. Richard Lindzen and James Holton developed a theory for the quasi-biennial oscillation (QBO) in two papers in the late 1960s and early1970s.[3] The model presented here is a simplified two-layer version of the Lindzen–Holton model, developed by Shigeo Yoden and James Holton.[4]

The basic principle behind this model, and all other models of the QBO, is as follows: Certain atmospheric waves carry momentum, in the same direction as their horizontal propagation, as they propagate upward through the atmosphere. This momentum is deposited in the mean flow at the levels where the waves dissipate. In other words, eastward-traveling waves carry eastward momentum, and westward-traveling waves carry westward momentum. Thus, waves can influence the mean (here in the sense of an

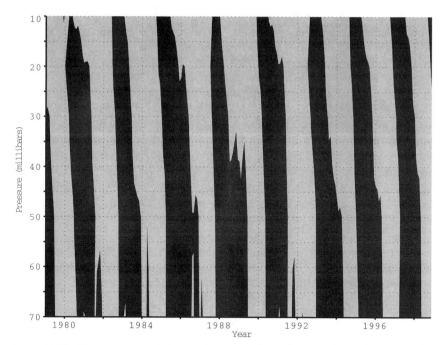

FIGURE 7.10   Twenty years of observed equatorial zonal winds. (Data from NCEP/ NCAR reanalysis project, Kalnay et al.[2])

average over longitude) winds. The mean winds, however, also influence the propagation of the waves. As the speed of the wave relative to the mean wind decreases, the vertical propagation of the wave is slowed, and it dissipates over a shorter vertical distance.

Imagine, for example, that we start with mean winds that are at rest and with a wave propagating from west to east. Where this wave is dissipated, it accelerates the mean winds toward the east. As the mean winds become more and more westerly, the wave dissipation and the deposition of eastward momentum occur lower and lower in the atmosphere. Thus, the interaction with a single wave gives rise to descending wave-driven winds. The oscillation of the QBO arises because the atmosphere supports both eastward- and westward-propagating waves. In the presence of westerly winds, the dissipation of westward-traveling waves is reduced, and these waves, therefore, reach greater altitudes before depositing their westward momentum. So, when the winds are westerly at lower altitudes, the acceleration is westward aloft, and vice versa. This causes the oscillation.

The model of the QBO (Figure 7.11) works in just this way. The winds at both levels, represented as stocks, are accelerated eastward by eastward-traveling waves, and westward by westward-traveling waves. The four flows, Eastward_high, Westward_high, Eastward_low, and Westward_low, represent these accelerations. The complicated expressions in these flows

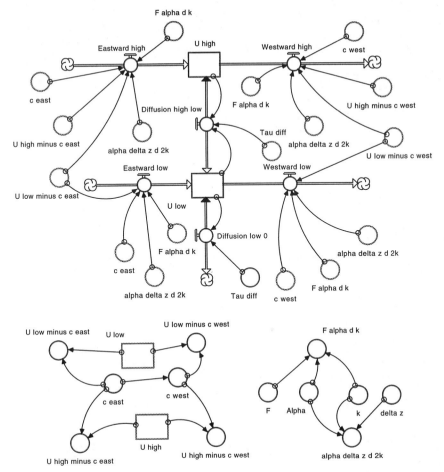

FIGURE 7.11   A model of the quasi-biennial oscillation (QBO).

capture the dependence of the wave absorption on the difference between the wave velocity and the wind. In addition, there is a diffusive exchange of momentum between the two levels, and the lower level exchanges momentum diffusively with lower atmospheric layers that are assumed to be at rest.

If the wind starts at rest, it will remain so forever, because the eastward and westward wave forces are in perfect balance. This equilibrium state is, however, unstable. A small perturbation of the winds in either level in either direction causes the oscillation to develop. Figure 7.12 shows the oscillation of the winds in this model. (Note that the abscissas in Figure 7.12 are displaced so that the upper-layer winds are shown above the lower-layer winds; the winds in both layers oscillate around zero. Also note that the units of wind speed and time are nondimensional.) While the shape of the oscillation is different in the two layers, the model is realistic in that the

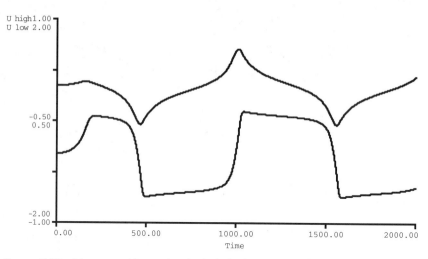

FIGURE 7.12   Upper- and lower-level winds in the QBO model.

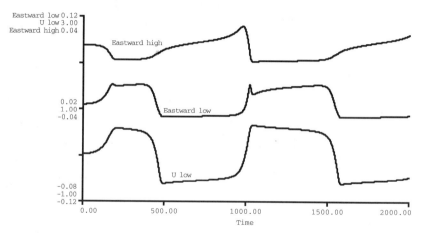

FIGURE 7.13   Lower-level winds and the eastward force on the upper- and lower-level winds in the QBO model.

upper-layer winds invariably lead those in the layer; the model captures the alternating descent of easterly and westerly winds.

That the dynamics of the model are as advertised is confirmed by examining the forces on the winds in relation to the winds themselves. Figure 7.13 shows the lower-layer winds (bottom curve); the eastward push of the waves on the lower layer, Eastward_low (middle curve); and the eastward push of the waves on the upper layer, Eastward_high (top curve). The positive feedback is evident in the lower layer, where stronger westerly winds cause more absorption of eastward traveling waves and a greater eastward push from those waves. The wave-shielding effect that is essential for the oscillation is seen by comparing the top and bottom curves. The eastward push on the upper layer nearly vanishes when the lower-layer winds are most westerly. In this phase of the oscillation, the eastward-traveling waves are all absorbed in the lower layer and do not make it to the upper layer. When the lower-layer winds are easterly, the eastward-traveling waves do reach the upper level and provide an eastward acceleration. The shielding of the upper layer from waves traveling in the same direction as the lower-layer winds acts, therefore, to increase the vertical shear in the winds. Through the diffusive connection between the layers, this provides, with a temporal lag, a negative feedback on the lower-layer winds.

If the strength of the wave forcing, F, is varied in this model, it is found that the period of the oscillation varies considerably with F, and further, that oscillations exist in the model only for F within a rather narrow range of values. Given that the period of the QBO in the atmosphere is quite stable over the time it has been observed, since the early 1950s, it suggests that many different waves are responsible for the atmospheric QBO and that their statistical properties, taken in aggregate, vary little over time. A completely periodic signal is, of course, perfectly predictable, so it would be useful indeed if the QBO were associated with climate variables for which predictions were desired. The QBO does, in fact, appear to have a weak influence on two climate variables of practical interest: the extent and severity of Antarctic ozone depletion in a given year (the so-called ozone hole), and the number and severity of Atlantic hurricanes.

7.3 Stommel's Model of the Thermohaline Circulation

The thermohaline circulation in the ocean is a planetary-scale flow driven by differences in the density of seawater at different locations. At a given pressure, the density of seawater depends on both its temperature and its salinity; cold water is denser than warm water, and salty water is denser than fresh water. Surface seawater becomes dense in high latitudes, where it loses heat to the atmosphere. If the surface water is cooled sufficiently, it can become denser than the surrounding water and sink convectively. This process, called deep-water formation, occurs primarily in the North Atlantic Ocean and around Antarctica; deep-water formation in the North Pacific is prevented by the lower salinities there. The formation of deep water in the North Atlantic drives a global "conveyor belt" circulation, wherein water that sinks in the North Atlantic spreads equatorward at depth, on into the South Atlantic around Antarctica, and into the Pacific and Indian basins. Outside of the regions of deep-water formation, there is compensating upwelling. Surface waters drift back into the Atlantic and northward through the Atlantic basin to replace the surface water lost by sinking. The northward flow of surface water in the Atlantic carries a great deal of heat. It is interesting to note that the Atlantic Ocean transports heat northward in both the Northern and the Southern Hemispheres, so that in the Southern Hemisphere it carries heat *toward* the equator.

The thermohaline circulation thus plays a huge role in the climate, especially in warming Europe, which is on the lee shore of the North Atlantic. Disruptions of the thermohaline circulation are implicated in rapid climate changes that have occurred in the past. An example is the sudden cooling called the Younger-Dryas cold event, that interrupted the warming trend at the end of the last ice age, about 12,000 years ago. It has also been suggested that the thermohaline circulation could collapse as a consequence of anthropogenic global warming, causing temporary cooling in Europe and a radical change in the marine environment. Most of the current generation of coupled atmosphere–ocean climate models predict that the strength of the conveyor-belt circulation will decrease significantly over the course of this century as global warming proceeds.

The possibility of a collapse of the thermohaline circulation, the so-called thermohaline catastrophe, suggests that the thermohaline circulation is not entirely stable. What is the nature of its instability? That the thermohaline circulation might, indeed, be unstable, in the sense that it could be flipped into a different and weaker flow, was first suggested in a beautiful paper by Henry Stommel.[5] He proposed the simple and elegant model presented here. There have been countless elaborations on Stommel's model, aiming at greater realism. A recent such elaboration is presented in the next section.

Stommel's model is shown in Figure 7.14. There are two ocean reservoirs, one at high latitudes, the other tropical. The temperatures and salinities of the reservoirs comprise the model's four stocks. Radiation and interactions with the atmosphere cool the northern reservoir, and the tropical reservoir is similarly warmed. For purposes of simplicity, these processes are represented by linear relaxations toward reference temperatures, T_cold and T_warm. For the most part, salt neither enters nor leaves the ocean, but freshwater does, by evaporation and precipitation. These processes tend to freshen the ocean in high latitudes and make it saltier in low latitudes (the atmosphere transports water vapor poleward; see Sections 2.7, 2.8, and 6.4). Again, these processes are represented as linear relaxations toward reference salinities, S_cold and S_warm.

The density of seawater in the ocean depends in a complicated nonlinear fashion on the temperature, salinity, and pressure. For simplicity, Stommel's model assumes a linear dependence on temperature and salinity. The flow of water between the reservoirs is assumed to be proportional to the difference in their densities—one can imagine two basins connected by a small-diameter tube near their bottoms, representing the abyssal flow, and a wider trough connecting them at the top, representing the surface flow. The trough ensures that the depth of water in the two basins is identical, so that the difference in pressure at the bottom, and thus the flow through the tube, is proportional to the density difference. These flows carry the properties, heat and salinity, of one ocean reservoir into the other.

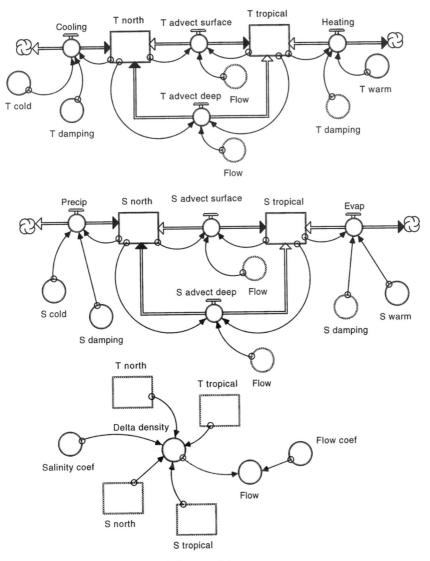

FIGURE 7.14   Stommel's thermohaline model.

This completes the structure of the model. There are, however, two criti-cal details on which its interesting behavior depends. First, consistent with observations that atmospheric processes modify the temperature of the upper ocean more rapidly than its salinity, the rate, S_damping, at which the salinity relaxes toward its reference values is several times slower than the rate, T_damping, at which the temperature relaxes. Second, consistent with the greater importance of salinity than temperature in determining the density of seawater, especially at low temperatures, the density dependence

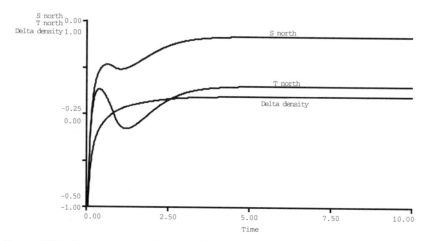

FIGURE 7.15   Temperature, salinity, and the north–south density contrast for a run of Stommel's model that equilibrates with sinking in the north.

on the salinity, Salinity_coef, is twice that on temperature. Note that, following Stommel's paper the model is nondimensionalized, so that all of these values are without units, and the constant relating changes in density to changes in temperature is assigned a value of 1.

As a result of these details, if the flow in the model is strong, the salinity in each reservoir will have little time to equilibrate with its reference value, and the difference in the temperatures of the reservoirs largely determines the difference in their densities. In this case, the cold northern water sinks, and the thermohaline circulation has the same sense, indicated by positive Flow in the model, that it does today in nature. If the flow is weak, however, the salinities have time to approach their reference values, and the salty water in the equatorial reservoir has the greater density and sinks. In this case, the circulation is in the opposite sense from that observed in today's ocean.

Do both of these regimes really exist in this model? If the model is started with temperatures and salinities equal to their reference values, Delta_density, the difference in density between the north and tropical reservoirs is initially negative, and Flow is likewise, with sinking in the south (Figure 7.15). The strong density difference, however, means that the flow is strong and the difference in salinity between the reservoirs is rapidly removed so that S_north approaches zero. (Note that, because the model maintains perfect antisymmetry between the north and the tropics, only one set of values need be plotted.) With the salinity difference removed, the temperature dominates the density, resulting in positive Flow with denser water (positive Delta_density) sinking in the north. To locate a regime in which the salinity dominates and sinking occurs in the tropics, it is necessary to start with a weak density difference. If the temperatures are initialized with their

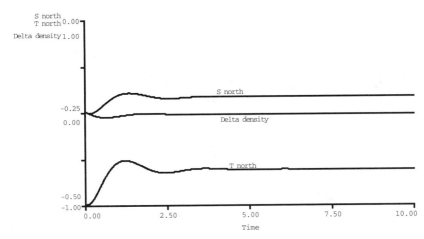

FIGURE 7.16   Temperature, salinity, and the north–south density contrast for a run of Stommel's model that equilibrates with weak sinking in the south.

reference values but the salinities start from one-half of their reference values, then the two reservoirs begin with equal densities and initially there is no flow. Figure 7.16 shows the results of this simulation. Flow remains sufficiently weak that the salinity is significantly lower in the northern reservoir; it remains well below zero. Even though the temperature is close to its reference value, the salinity dominates the density, so Delta_density and Flow are both slightly negative. Note that in this case, with sinking in the south, the temperature remains close to its reference value of −0.5, while when there is sinking in the north and strong flow, T_north equilibrates at around −0.2. The weak and reversed thermohaline circulation leads to a much colder climate in northern high latitudes.

A change in the thermohaline circulation can change the climate, but what happens to the thermohaline circulation as the climate changes? This model does not care about the absolute values of the reference temperatures, only the difference between them. Almost all models of global warming, however, project a reduction in the pole-to-equator difference as anthropogenic global warming proceeds, and, in fact, such a reduction is already being observed. To simulate this in Stommel's model, a slow linear increase (.002 units/year) with time is introduced to T_cold, and an identical decrease with time to T_warm. The model starts from equilibrated conditions with sinking in the north.

The results of this "climate change" experiment are shown in Figure 7.17. Initially the temperature, T_cold, changes little as T_cold warms. The reduction in Cooling is nearly balanced by a similar reduction in the warming influence of the overturning circulation, T_advect_surface – T_advect_deep, as Delta_density decreases. To the extent that this model can be viewed as a simulation of the North Atlantic Ocean, it suggests the possibility that

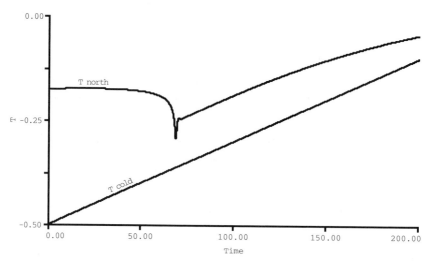

FIGURE 7.17   Northern and cold reservoir temperatures in a "climate change" experiment with Stommel's model.

temperatures in the high-latitude North Atlantic might be relatively insensitive to global warming, and this is, in fact, what the observed temperature record shows. As T_cold warms to about −0.4, T_north begins to *decrease,* initially slowly, then sharply. For values of T_cold higher than about −0.35, T_north tracks the warming in T_cold. This transition marks the point at which the "thermohaline catastrophe" occurs, and the circulation reverses from a strong temperature-driven sinking in the north to weak salinity-driven sinking in the tropics. This is readily confirmed by plotting the variations in Flow, or Delta_density. The initial sharp cooling of T_north at this transition suggests the possibility of a seemingly paradoxical outcome of global warming, that temperatures in the mid- to high-latitude North Atlantic and possibly downwind over Europe could decrease at the same time that the earth as a whole, especially elsewhere at high latitudes, is warming.

7.4 A More Realistic Thermohaline Model

Despite its conceptual beauty, the previous model has some serious shortcomings as a representation of the real world. The thermohaline circulation in the Atlantic is approximately a single cell with sinking in high northern latitudes and upwelling in mid to high southern latitudes, so that there is a northward drift of surface waters through most of the Northern and Southern Hemispheres. This circulation is driven by the difference in density between waters in high northern and southern latitudes, a density difference that is primarily due to the greater salinity of the water in the North

Atlantic. Also, it is hard to justify, on grounds other than conceptual simplicity, the use of a linear relaxation for salinity. The amounts of salt and water in the ocean are roughly conserved, and salinity is modified at any location in the surface water through the gain and loss of fresh water by precipitation and evaporation. The water that rains on one part of the ocean must ultimately have evaporated from another part, so that salinity differences between regions are maintained by the flux of fresh water through the atmosphere.

Recently, Scott, Marotzke, and Stone[6] proposed a model that captures these features of the real world. This model is shown in Figure 7.18. The structure is similar to that of Stommel's model, with two key differences: First, three ocean reservoirs represent the Southern Hemisphere, the Northern Hemisphere, and the Tropics. Second, the linear relaxation that governed the salinity in Stommel's model is replaced by fluxes of salinity from the high-latitude reservoirs to the tropics—this being opposite in direction from the atmospheric transport of fresh water. The model is now closed with respect to salinity; the total salinity cannot change. With the parameters in this model set to their nominal values, it produces an equilibrium circulation that mimics the real world in that northern temperatures and salinities are both significantly higher than those in the Southern Hemisphere, and there is sinking in the north. Note that this model uses physical units. These are standard, except perhaps for salinity, which is measured in psu. For present purposes, these may be considered parts per thousand, or grams of salt per kilogram of seawater.

This model has a number of interesting and subtle features. Because the salinity drives the circulation, the behavior is sensitive to the magnitudes of the salinity fluxes, S_flux_s_tr and S_flux_n_tr. For a standard situation, in which water sinks in the Northern Hemisphere, however, the flux of salinity from the north to the tropics cannot directly influence the strength of the circulation. Only the salinity flux in the Southern Hemisphere effects the difference in salinity, and thus density, between the north and south. Figure 7.19 shows the results of an experiment in which the flux of salinity from the south to the tropics is increased linearly so as to double over 10,000 years. This implies increasing evaporation in the tropics and increasing rainfall in the southern extratropics. As the water in the south becomes less salty and less dense, the density difference between the north and the south increases and so does the strength of the circulation. The stronger circulation warms both hemispheres in high latitudes, but especially the north. Thus, this model predicts that a variation in the atmospheric transport of water vapor from the tropics to the *Southern* Hemisphere could have a significant effect on the climate of high *northern* latitudes. This is an intriguing possibility, if one that is difficult to test.

While the atmospheric flux of salinity from the tropics to the north cannot directly influence the strength of the circulation, it can affect the stability of that circulation. For a wide range of parameter settings, this model, like Stommel's, possesses two stable equilibrium states, one with sinking in

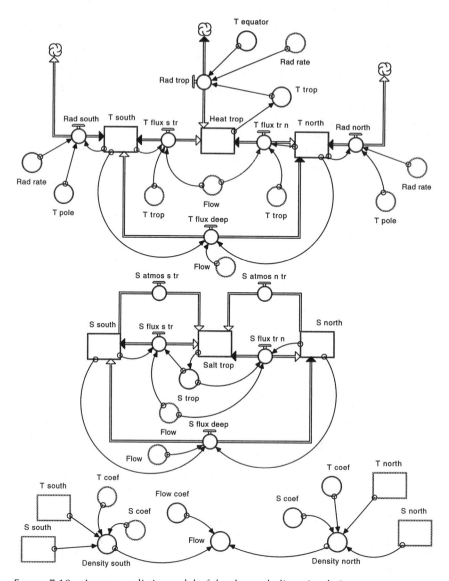

FIGURE 7.18   A more realistic model of the thermohaline circulation.

the north, the other with sinking in the south. It has been suggested that the Younger-Dryas cold event was triggered by a flux of fresh water into the North Atlantic during the rapid retreat of the North American icesheet. The stability of the thermohaline circulation in this model under such pulses of fresh water can be investigated by perturbing the model with reductions of varying strength in the initial salinity of the North Atlantic and seeing how great a perturbation is needed to reverse the circulation. Figure 7.20 shows

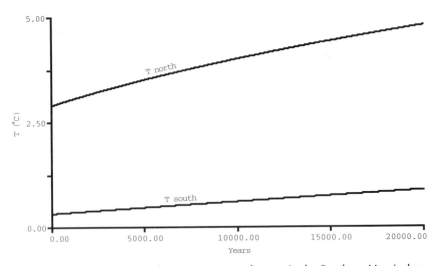

FIGURE 7.19   The response of temperatures to changes in the Southern Hemisphere salinity flux.

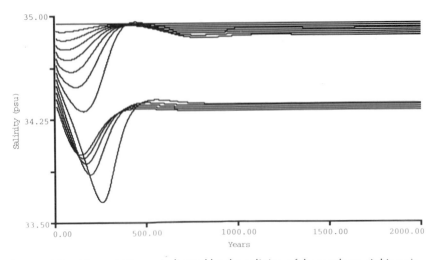

FIGURE 7.20   The stability, as indicated by the salinity, of the northern-sinking circulation under initial perturbations of the salinity.

the Northern Hemisphere salinity in experiments that start from the standard equilibrium state but include increasingly strong reductions in the Northern Hemisphere salinity. The top curve is unperturbed, and moving down the figure, each curve corresponds to an additional 0.05 psu reduction in the initial salinity. For this experiment, a reduction of 0.4 psu is required to cause a reversal of the thermohaline circulation. When the experiment is repeated, however, with the Northern Hemisphere's atmospheric

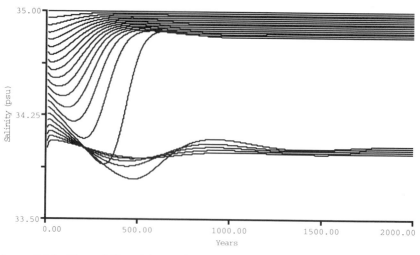

FIGURE 7.21   The stability of the northern-sinking circulation with a reduced Northern Hemisphere salinity flux.

salinity flux reduced by a factor of 3, a reduction of 0.8 psu is required (Figure 7.21). Thus, it is seen that increased atmospheric transport of water vapor from the tropics to northern high latitudes—more rain in the northern extratropics—reduces the stability of the thermohaline circulation. As the climate warms, and the hydrologic cycle becomes more vigorous, this result suggests that the thermohaline circulation could become more subject to catastrophic reversals.

7.5 Icesheet Model

Over the past 2 million years, great sheets of ice, kilometers thick, have repeatedly advanced southward over North America, Europe, and Asia. Here, we present a simple model of the dynamics of an icesheet and consider its implications for some features of the ice ages. The model, due to Johannes Weertman,[7] is depicted in the schematic in Figure 7.22. The STELLA model is shown in Figure 7.23. There is a single stock for the ice volume, with units of cross-sectional area, square meters. The ice volume increases by Accumulation and decreases by Ablation. The ice is assumed to flow as a plastic solid, which gives it a cross-sectional profile of a parabola laid on its side. Ice poleward of the highest point flows northward; ice to the south of the highest point flows southward. Ice accumulates on the surface of the icesheet where the temperature at the surface of the icesheet is below freezing—the accumulation zone—and is lost from the surface in those regions where the temperature is above freezing—the ablation zone.

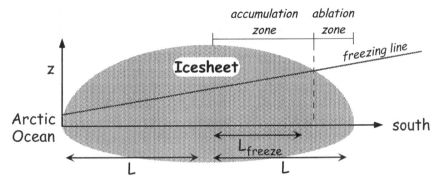

FIGURE 7.22   A schematic of a model icesheet.

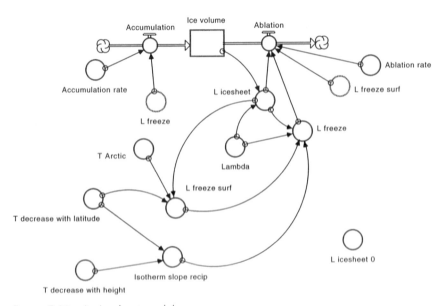

FIGURE 7.23   An icesheet model.

The geometry, as shown in the figure, is crucial for the behavior of the icesheet. First, the icesheet is assumed to be bounded to the north by the Arctic Ocean. Where ice flows out into the ocean, it calves into icebergs. North of its highest point, it is assumed that the accumulation zone is always greater in area than the ablation zone, so that only the shore of the Arctic Ocean limits the ice extent. Because temperatures increase southward, but decrease upward, the freezing line slopes upward to the south. Thus, the southward extent of the ice is limited when it comes to an equilibrium in which the area of the accumulation zone south of the highest point on the icesheet is equal to the area of the ablation zone. Because the earth beneath the icesheet is pushed downward by the great weight of the

ice, the icesheet extends down below sea level. It is assumed that this adjustment of the earth's crust is isostatic; i.e., the icesheet effectively floats on
the underlying rock, so that the vertically integrated mass of material, rock,
and ice is everywhere the same. Then, if the density of the underlying rock
is three times that of the ice, the depth of the icesheet below sea level is
one-half of its elevation.

If some region of the icesheet is above the freezing line, the icesheet
comes to a dynamic equilibrium in which the volume of ice gained in the
accumulation zone balances that lost in the ablation zone. It is important to
realize that in this equilibrium, while the ice edge is neither advancing nor
retreating, there is a constant flow of ice within the sheet from the accumulation to the ablation zone. In Weertman's model, it is assumed that when
the entire icesheet is in the ablation zone (in other words, when the entire
surface is above freezing), the decay of the icesheet is then increased to
three times the normal ablation rate. Otherwise, the rate of ice ablation in
the ablation zone is assumed to be 2.75 times greater than the rate of accumulation in the accumulation zone, which in turn is set to 1.2 m/year.

The position of the freezing line is set by the temperature at the shore of
the Arctic Ocean, T_Arctic, and by the rates of southward temperature increase and upward temperature decrease. Because the latter is much larger
than the former, the freezing line is nearly horizontal. This fact, together
with the shape of the icesheet, is responsible for much of the interesting
behavior displayed by this model. Very roughly, we can imagine the shape
of the icesheet as being something like that of a drop of water resting on a
sheet of glass. When the volume of the drop is small, it is nearly a hemisphere, but as we add water, it spreads out and gets flatter and flatter on
top. Similarly, the icesheet gets flatter and flatter in its cross-sectional profile
as its volume increases. For its size, a smaller icesheet sticks higher into the
air, where the temperatures are colder. Even without any southward increase in temperature, if the temperature at the surface is above freezing,
the icesheet can grow until it flattens sufficiently that enough of its surface
is below the freezing line to achieve equilibrium, although it may become
very large before this happens.

Figure 7.24 shows the results of a series of runs with T_Arctic set to 3°C.
Each starts with a different initial ice extent. When the icesheet is too small,
its top is below the freezing line, and it rapidly disappears. When the icesheet is sufficiently large, however, it can grow, and it grows rapidly to a
final size that is independent of its initial size. The icesheet is a system,
then, with two stable equilibria, one at zero extent and the second at a
much larger size, a horizontal extent of about 1000 km (twice the value of
L_icesheet) in this case. These equilibria can be explored by saving in a
table and then graphing the accumulation and ablation rates for icesheets
with a range of sizes. The results are shown in Figure 7.25. When the icesheet is very small, there is no accumulation zone, and the ablation is rapid.
Even when the top of the icesheet is above the freezing line, ablation exceeds accumulation until the icesheet reaches a critical size, L_icesheet ~

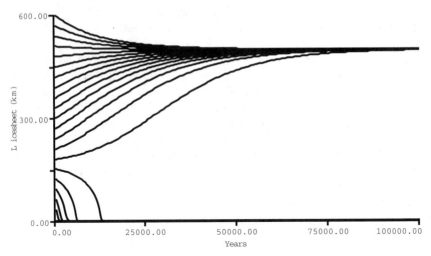

FIGURE 7.24   Results of runs of the icesheet model with different initial ice extents.

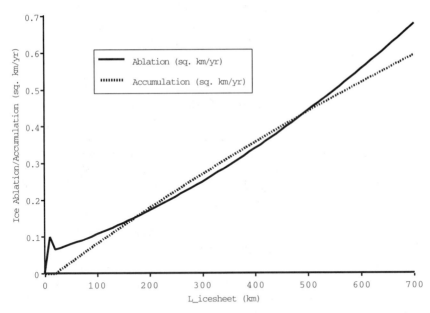

FIGURE 7.25   Rates of ice accumulation and ablation versus the ice extent, L_icesheet.

150 km). At this size, the ablation and accumulation are equal, but this is not the ice extent reached by the model when it is run. As in the climate model with albedo feedback (Section 2.3), this is an unstable equilibrium. If the icesheet is slightly larger than this value, the accumulation is greater than the ablation, so the icesheet continues to grow. The crossing of the

two curves at the larger value of L_icesheet, around 500 km, is a stable equilibrium, and this is the one reached by the model.

How does the equilibrium ice extent change with the temperature? As long as the temperature at sea level on the shore of the Arctic Ocean, T_Arctic, is above freezing, it is clear that the absence of ice is a stable equilibrium. When the model is run with different values of T_Arctic, starting from an initial condition with a large icesheet, it is found that there is a second equilibrium with a substantial icesheet for all values of T_Arctic less than or equal to 3°C. Thus, for values of T_Arctic between 0 and 3°C, there are two stable equilibria, one with a substantial icesheet, and one with no ice. Which one appears will depend on the previous state of the system. When T_Arctic is cooling from a value above 3°C, there will be no ice until it cools to the freezing point, but if it is warming from below freezing, an icesheet will persist until it warms to 3°C.

Application to the Ice Ages

According to the astronomical, or Milankovitch, theory of the ice ages, the advances and retreats of icesheets during the past 2 million years were driven by a periodic modulation in the amount of sunlight reaching Northern high latitudes during summer. The source of the modulation is the precession of the equinoxes, the variation of the timing of Earth's closest approach to the sun relative to the seasons. If the earth is at perihelion (closest approach to the sun) in northern summer, then the intensity of solar radiation during that season is enhanced, while the opposite is true if northern summer corresponds to aphelion (greatest distance from the sun). The seasonal timing of perihelion varies with a period of about 21,000 years. In addition to the precession of the equinoxes, the amount of sunlight reaching high latitudes is also modulated by variations in the earth's axial tilt, but these are set aside here in the interest of simplicity.

Why is northern summer the key season? The thinking, originally due to the mathematician Milutin Milankovitch, is that winters are always cold enough for snow to accumulate in high northern latitudes. Therefore, whether or not an icesheet can form and grow depends on the extent to which the ice ablates during the summer. It is northern summer that matters, because in the Southern Hemisphere, an ice-covered continent is surrounded by ocean, and there is no room for icesheets to grow.

Milankovitch's theory, when tested against climatic records preserved in deep-sea sediments and coral reefs, has proven successful, and it is widely accepted. A puzzle remains, however. The dominant variation in these paleoclimatic records is not at a period of 21,000 years, but rather at a period close to 100,000 years. The eccentricity of the earth's orbit about the sun, how much the elliptical orbit deviates from a circle, varies on about

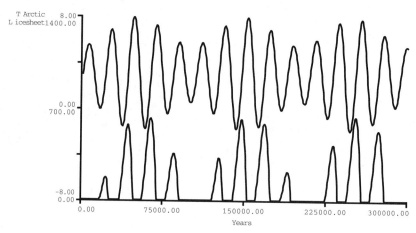

FıGURE 7.26   The Arctic temperature, T_Arctic (top curve), and the ice extent, L_icesheet (bottom curve), when the icesheet model is run with Milankovitch-like variations in T_Arctic.

this period. The more eccentric the earth's orbit, the greater the effect of precession on the summertime insolation. Simply put, the summertime insolation should be like a sine wave with a period of 21,000 years, but with an amplitude that should vary with a period of 100,000 years. If the growth and decay of icesheets followed the summer insolation, there would be ice ages spaced 21,000 years apart, the strengths of which would vary with a 100,000-year cycle. This is not what is seen in the climatic record, however. Rather, it is seen that the ice extent itself varies most strongly on the 100,000-year period.

We can apply our icesheet model to this problem. T_Arctic is now taken to be representative of summertime conditions, and it is varied periodically about 3°C with a period of 21,000 years. The amplitude of this 21,000-year cycle is itself varied between 2.5 and 5°C with a period of 100,000 years. Figure 7.26 shows the time history of both T_Arctic (top curve) and L_icesheet (bottom curve) over 3 such 100,000-year cycles. T_Arctic is a sinusoid with a periodically varying amplitude, as specified. L_icesheet, however, shows periods of ice activity that are separated by 100,000 years. Essentially, the nonlinearity of the response of the icesheet to variations in T_Arctic allows it to rectify the amplitude-modulated signal and extract the 100,000-year period, in much the same way as the circuits of an AM radio extract an audio signal from a radio-frequency carrier wave. Note that because the icesheet takes thousands of years to grow, the coldest periods correspond not to the greatest ice extent but to the time of most rapid glacial advance. This is quite in accordance with what is observed in the paleoclimatic record.

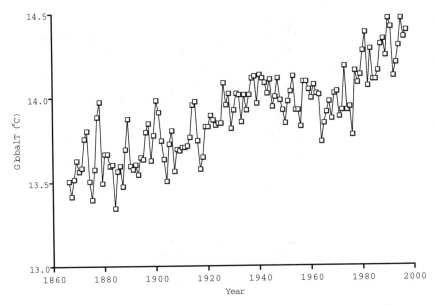

FIGURE 7.27   Globally averaged surface temperatures for the past century. (From Hansen et al.[8])

7.6 Upwelling–Diffusion Climate Model

Even a cursory examination of the record of globally averaged temperatures over the past century (Figure 7.27)* reveals that the temperature has increased, but it has not increased smoothly. While it is widely believed that the warming trend results from anthropogenic increases in atmospheric concentrations of greenhouse gases, especially carbon dioxide, these increases have been smooth and monotonic and cannot explain the rapidity of the warming before 1940 or the slight cooling between 1940 and 1970. It is possible that some other external agent, such as variations in solar luminosity, drove these variations in temperature, but it is reasonable to ask whether internal variability within the climate system might be responsible.

Because the heat capacity of the deep ocean is very much greater than that of the upper ocean and atmosphere, it is conceivable that fluctuations in the climate measured, as in Figure 7.27, at the surface, could result from exchanges of heat between the upper ocean/atmosphere and the deep ocean. How might this work? The meridional overturning circulation in the

*This record is derived from surface meteorological stations, with urban heat-island effects removed. It is the product of the Goddard Institute for Space Studies[8] and is available on the World Wide Web: http://www.giss.nasa.gov/data/update/gistemp/

ocean, modeled earlier in this chapter, is driven by a process called deep-water formation. In high latitudes, in the North Atlantic Ocean and at some locations around Antarctica, surface waters become sufficiently dense to sink through the water column. Particularly in the North Atlantic, this process is episodic. It occurs in winter but only in limited regions and at times when strong winds and cold air remove heat effectively from the surface waters. The formation of deep water is, therefore, influenced by the chaotic dynamics of the subpolar atmosphere, and it is reasonable to expect that the amount of deep water formed will vary substantially from winter to winter. Such variability has recently been confirmed by oceanographic observations.

Water is conserved, so the formation of deep water must be balanced, as in the models in Sections 7.3 and 7.4, by water upwelling elsewhere in the ocean. It turns out that this upwelling water is warmer than the water that sinks in high latitudes, so the process of deep-water formation, taken as a whole, carries heat from the deep ocean to the surface. If the rate of deep-water formation temporarily increases, the surface will warm at the expense of the deep ocean, while if it decreases, the surface will cool.

The simplest possible model that can represent this process is shown in Figure 7.28. The stocks represent the heat content (heat capacity times temperature) of the upper and deep ocean. Heat is transferred from the upper to deep ocean by deep-water formation, while heat is transferred from the deep to the upper ocean through upwelling. A key point is that while the upwelling water is at a temperature characteristic of the entire deep ocean, the sinking water is *not* at the temperature of the surface ocean; rather, it is at the colder temperature, T_sink, required for it to sink to depth. Diffusion also transfers heat between the upper and deep ocean. This is a downward transfer, because the surface is warmer than the deep. The result of this diffusion is that the deep water is warmer than T_sink. Thus, the combination of upwelling and sinking transfers heat from the deep to the surface. In equilibrium, this loss from the deep is made up by the downward diffusive flux of heat.

The remainder of the model is a very simple representation of the effects of radiation. The surface is assumed to have a radiative equilibrium temperature toward which the surface temperature is damped at a rate that depends on the climate sensitivity, or Gain. The nominal value of $3 \text{ W m}^{-2}\text{K}^{-1}$ corresponds to a case with no positive feedback. If Gain is increased, the strength of this negative feedback decreases proportionally.

With this model, we can first consider the effects of random fluctuations in the amount of deep-water formation. It is assumed that in any year the amount of deep water formed can take on any value between zero and twice the nominal amount. Note that the nominal rate of deep-water formation is such that would flush the deep ocean in 500 years, while diffusion acts on the deep ocean on a timescale of 3000 years. Figure 7.29 shows three 1000-year simulations with randomly fluctuating deep-water formation (Random_switch set to 1) and with the Gain set to 1, 2, and 4

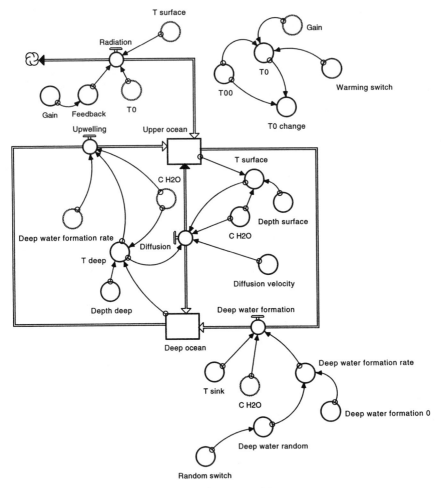

FIGURE 7.28   An upwelling–diffusion climate model.

(from the bottom curve to the top; note that for purposes of display, the temperature is offset by one degree for each change in Gain). As Gain, or the climate sensitivity, is increased, both the amplitude and the duration of fluctuations in the temperature become greater, much as was the case when the size of the hole was reduced in the leaky bucket model (Section 1.2). This is the first important result from this model: *The more sensitive the climate is to external forcing, the more robust and persistent are its internal fluctuations.*

How does the model respond to changes in external forcing? In this model, all external forcing, be it changes in the luminosity of the sun or in the greenhouse properties of the atmosphere, are represented by the temperature in equilibrium, T0. If the climate is more sensitive than would be

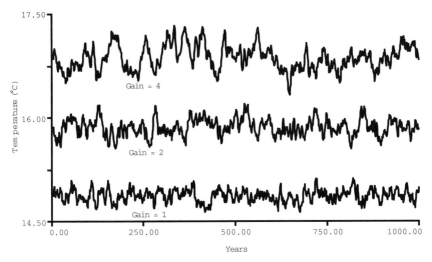

FIGURE 7.29   Surface temperatures in the upwelling–diffusion model with randomly varying rates of deep-water formation, for three different values of the climate sensitivity. The top curve corresponds to the most sensitive climate.

the case for an earth with no atmosphere, the gain is increased, and the response of T0 to a certain radiative perturbation must be increased accordingly. Here, T0 is increased over the course of 100 years in a way that roughly approximates the anthropogenic increase in greenhouse-gas concentrations during the 20th century (Warming_switch set to 1). The increase in the equilibrium temperature at the end of 100 years is given by Gain in units of degrees Celsius. Performing a sensitivity experiment as before, the equilibrium warmings at the end of the century are 1, 2, and 4°C. Figure 7.30a and b show the changes in the temperatures, T_surface, and the equilibrium temperatures, T0, for the three values of Gain. These are results for 100-year simulations, with no random variations in the rate of deep-water formation (Random_switch set to 0). Because of its finite heat capacity, the temperature lags behind its radiative equilibrium value. This lag is greater when the climate is more sensitive. For a transient warming, therefore, the more sluggish response to the change in equilibrium temperature partially compensates for its greater value. This is the second important result from this model: *The greater the sensitivity of the climate, the slower its approach to its radiative equilibrium temperature.*

To simulate the behavior of the climate over the past century, we need to include both increasing radiative forcing and year-to-year variations in deep-water formation. To explore the range of possible outcomes, an ensemble of experiments is run in each case, using identical parameters but different sequences of the random numbers that determine the deep-water formation. Figure 7.31a to c shows five-member ensembles of 100-year runs, in each case with (Warming-switch = 1) and without (Warming_switch = 0)

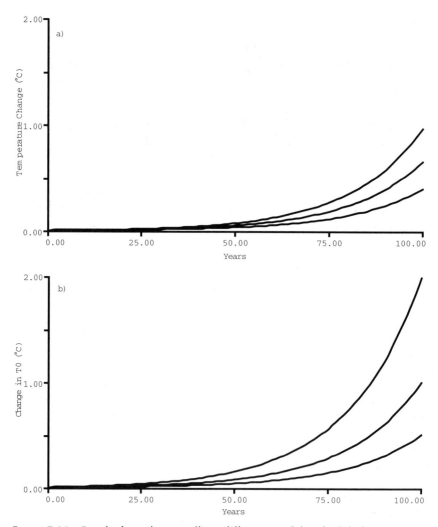

FIGURE 7.30   Results from the upwelling–diffusion model with global warming and without random variations in deep-water formation: (a) change in temperature; (b) change in equilibrium temperature, T0.

"global warming" (increasing values of T0), for Gain equal to 1, 2, and 4. The two ensembles with and without global warming clearly separate at the time when the "signal" of global warming clearly emerges from the "noise" of internal variability. At the highest value of Gain, this happens only about a decade sooner, at around year 70, than for the lowest value. The fact that the equilibrium warming for Gain = 4 is four times as great as for Gain = 1 is hidden in the delayed response and the greater internal variability in the Gain 4 case. This then leads to the final important result from this model: *In the early stages of global warming, it is difficult, if not impossible,*

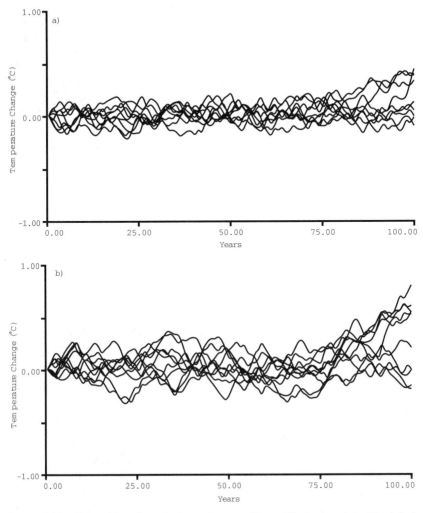

FIGURE 7.31   Ensembles of results from the upwelling–diffusion model with global warming and randomly varying deep-water formation, for different values of the climate sensitivity: (a) Gain = 1; (b) Gain = 2; (*continued*)

to estimate the sensitivity of the climate system by examining the observed temperature record.

Two final points in regard to this model: First, with increasing T0, it is not hard to produce a temperature record that qualitatively resembles the observed temperatures during the 20th century, though it is, in fact, quite hard to do so if T0 remains fixed. This is one approach that is being employed with coupled atmosphere–ocean GCMs to address the issue of detecting global warming.[9] Ensembles of integrations are run with and without increasing greenhouse gases. Many of the integrations that include increasing

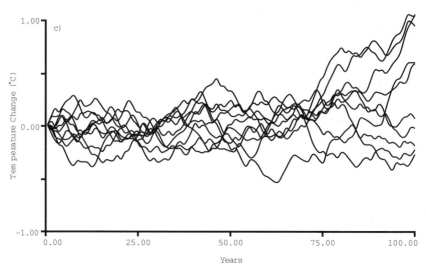

FIGURE 7.31   (*continued*) (c) Gain = 4.

greenhouse gases produce temperature records similar to what has been observed, especially in regard to the rapid warming over the past three decades, but few if any do so if the greenhouse gases in the model are not increased. To the extent that these models adequately represent the internal variability in the climate system, it is thus highly unlikely that the observed recent warming is other than the signal of global warming.

A second point relates to one of the most serious likely consequences of global warming, the rise in sea level. Sea level can increase locally for many reasons involving geology and hydrology, but global increases can result only from melting ice on land, either mountain glaciers or the great ice-sheets in Antarctica and Greenland, or from the thermal expansion of sea-water as the oceans warm. The deep ocean, however, warms very slowly, so even if the surface warming were to cease today, it would take many decades before the full increase in sea level is realized.

Problems

7.1 How robust are the self-sustaining oscillations in the El Niño model (Section 7.1)? For example, under what range of temperatures for the West Pacific, T_warmpool, are self-sustaining oscillations obtained? Are your results consistent with the observation, based on paleoclimatic evidence, that El Niño oscillations have been absent in some past periods? Are the oscillations in the stochastically forced model more or less robust than the self-sustaining oscillations?

7.2 Modify the QBO model to allow different strengths for the eastward- and westward-propagating waves. What happens to the oscillations as the asymmetry between the wave strengths is increased? Can you explain this result?

7.3 A general property of nonlinear systems such as Stommel's thermohaline model (Section 7.3) or the icesheet model (Section 7.5) is that their equilibrium state at any time depends not only on the present conditions but on their history as well. In other words, they exhibit *hysteresis*. In the case of Stommel's model, this can be explored by setting the cold reservoir temperature, T_cold, to slowly increase from -1.5 to 0, and then decrease slowly over the same range. To observe the hysteresis, use the "Scatter" feature to plot T_north against T_cold. Under what range of values of T_cold is the behavior of the model dependent on past conditions? Can hysteresis be found by varying model parameters other than T_cold?

Hysteresis can be observed in the icesheet model by slowly increasing and decreasing T_Arctic. Over what range of Arctic temperatures does the equilibrium extent of the icesheet depend on its history?

7.4 In the upwelling diffusion climate model (Section 7.6), what happens if at some time deep-water formation ceases? This experiment is most easily carried out with global warming and random fluctuations in the rate of deep-water formation turned off (Warming_switch = Random_switch = 0). Then Deep_water_formation_0 can be set to 0 after a certain number of years. If global warming does indeed lead to weakening of the conveyor belt, is this a positive or negative feedback for climate change?

7.5 A more realistic version of the upwelling diffusion model (Section 7.6) describes the ocean with several layers, rather than two. Build this more realistic model using five, or if you are ambitious, ten, subsurface layers in the ocean. As before, assume that all the deep water from the surface goes to the deepest ocean layer. Diffusion of heat downward between adjacent layers is proportional to the difference in their temperatures and the Diffusion_velocity. Heat is carried by upwelling from each layer to the layer above at a rate proportional to the temperature of the lower layer. In all cases, be careful about heat capacities and dimensions! What do you expect the equilibrium ocean temperature profile will look like? Run the model to steady state for a case with no global warming and no variations in the rate of deep-water formation (Warming_switch = Random_switch = 0), and check your prediction. Then repeat the global warming and random forcing experiments from Section 7.4. Are the results essentially the same as in the simpler model? How do they differ?

Further Reading

The Oceans and Climate, by Grant R. Bigg (1996, Cambridge University Press, 266 pp.), offers readable discussions of El Niño and the role of the thermohaline circulation in climate. For a detailed understanding of Stommel's thermohaline model itself, one can do no better than reading his very lucid original paper.[5] Observations and dynamics of the QBO are described in detail in *Middle Atmosphere Dynamics,* by David G. Andrews, James R. Holton, and Conway B. Leovy (1987, Academic Press, 489 pp.). A very readable historical and scientific review of the Milankovitch theory of ice ages is *Ice Age, Solving the Mystery,* by John Imbrie and Katherine Palmer Imbrie (1979, Enslow Publishers, 224 pp.), while Dennis L. Hartmann's *Global Physical Climatology,* (1994, Academic Press, 411 pp.), provides a more mathematical treatment. A review of the scientific and policy issues surrounding global warming is provided by *Global Warming: the Complete Briefing (Second Edition),* by John Houghton (1997, Cambridge University Press, 251 pp.) A deeper look at models, of which that described in Section 7.5 is the simplest example, is *An Introduction to Simple Climate Models Used in the IPCC Second Assessment Report—IPCC Technical Paper II,* edited by John T. Houghton, L. Gyvan Meira Filho, David J. Griggs, and Kathy Maskell, (1997, Intergovernmental Panel on Climate Change, 47 pp.).

References

1. Jin, F. F., 1996: Tropical ocean atmosphere interaction, the Pacific cold tongue, and the El Niño–southern oscillation, *Science,* **274,** 76–78.
2. Kalnay, E., *et al.,* 1996: The NCEP/NCAR reanalysis project. *Bulletin of the American Meteorological Society,* **77,** 437–471. Data available online at http://www.cdc.noaa.gov/
3. Lindzen, R. S., and J. R. Holton, 1968: A theory of the quasi-biennial oscillation. *Journal of the Atmospheric Sciences,* **25,** 1095–1107. Holton, J. R., and R. S. Lindzen, 1972: An updated theory for the quasi-bienncial cycle of the tropical stratosphere. *Journal of the Atmospheric Sciences,* **29,** 1976–1080.
4. Yoden, S., and J. R. Holton, 1988: A new look at equatorial quasi-biennial oscillation models. *Journal of the Atmospheric Sciences,* **45,** 2703–2717.
5. Stommel, H., 1961: Thermohaline convection with two stable regimes of flow. *Tellus,* **13,** 224–229.
6. Scott, J. R., J. Marotzke, and P. H. Stone, 1999: Interhemispheric thermohaline circulation in a coupled box model. *Journal of Physical Oceanography,* **29,** 351–365.
7. Weertman, J., 1976: Milankovitch solar radiation variations and ice age ice sheet sizes. *Nature,* **261,** 17–20.

8. Hansen, J., R. Ruedy, J. Glascoe, and M. Sato, 1999: GISS analysis of surface temperature change. *Journal of Geophysical Research,* **104,** 30997–31022). Data is available online at http://www.giss.nasa.gov/data/update/gistemp/

9. Knutson, T. R., T. L. Delworth, K. W. Dixon, and R. J. Stouffer, 1999: Model assessment of regional surface temperature trends (1949–1997). *Journal of Geophysical Research,* **104,** 30,981–30,996.

Appendix

A.1 System Requirements
    A.1.1. Macintosh
        Macintosh Minimum Requirements
            MacOS 7.1
            68040 Processor
            16 MB RAM
            35 MB Hard Disk Space
            Color Display running at least 256 colors.
        Macintosh Recommended Requirements
            120 MHz PowerPC or better
            MacOS 8.1 or higher
            32 MB RAM or higher
            35 MB Hard Disk Space
            Color Display running 'thousands of colors' or more
            QuickTime™ 3.0 or higher
    A.1.2 Windows
        Windows Minimum Requirements
            Windows 95, 98, Millennium Edition, NT 4.0 or Windows 2000
            Pentium-Class Processor
            16 MB RAM
            30 MB Hard Disk Space
            VGA Display with at least 256 colors
        Windows Recommended Requirements
            Windows 95, 98, Millennium Edition, NT 4.0 or Windows 2000
            233 MHz Pentium-Class Processor or better
            64 MB RAM
            30 MB Hard Disk Space
            SVGA Display with 16 bit color or better
            SoundBlaster or compatible sound card
            QuickTime™ 3.0 or higher

A.2 Quick Help Guide*

A.2.1 Overview of STELLA 6.0 Operating Environment

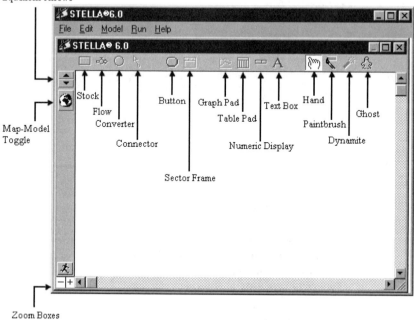

*By High Performance Systems, Inc.

A.2.2 Drawing an Inflow to a Stock

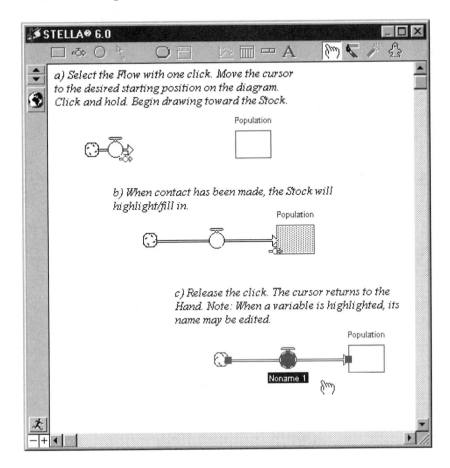

A.2.3 Drawing an Outflow from a Stock

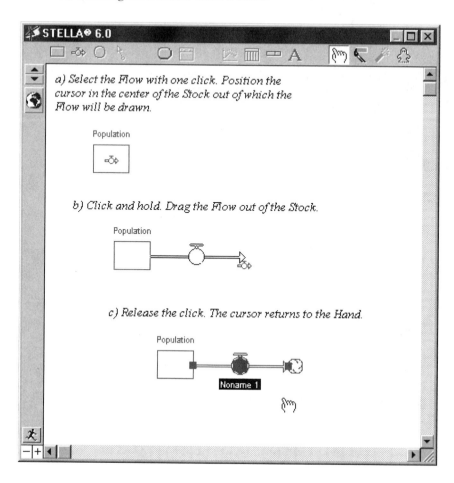

A.2.4 Replacing a Cloud with a Stock

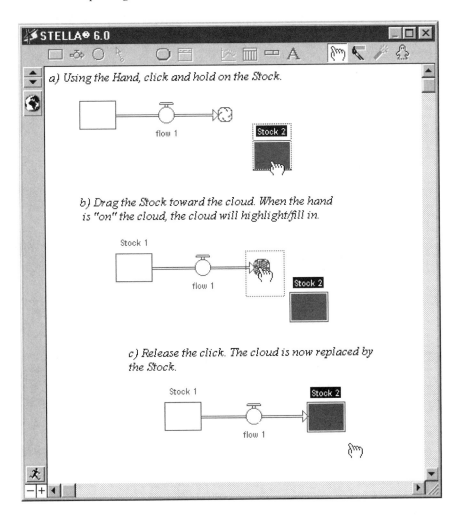

A.2.5 Bending Flow Pipes

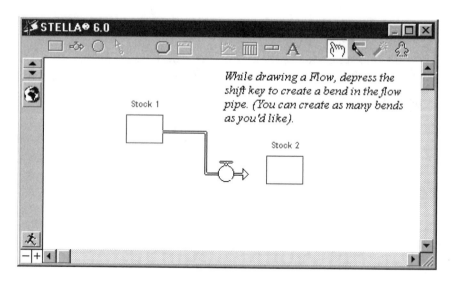

A.2.6 Repositioning Flow Pipes

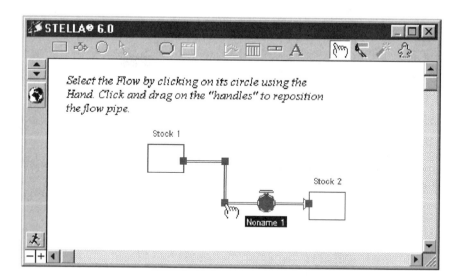

A.2.7 Reversing Direction of a Flow

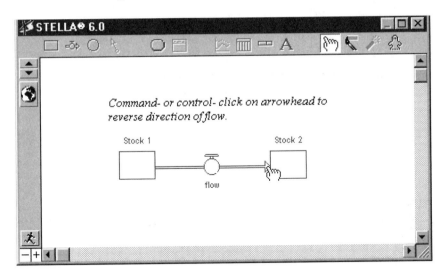

A.2.8 Flow Define Dialogue—Builtins

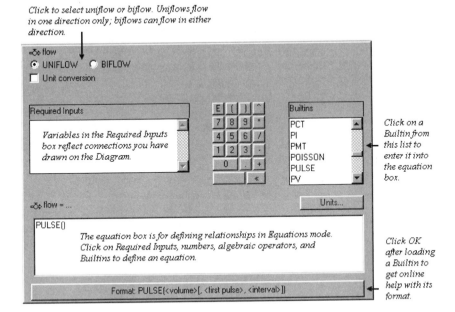

A.2.9 Moving Variable Names

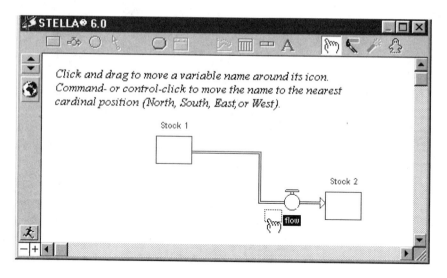

A.2.10 Drawing Connectors

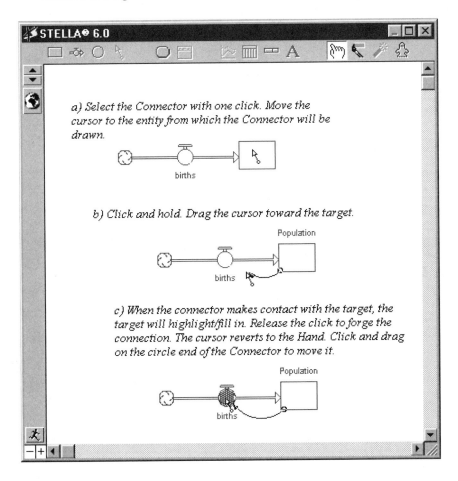

a) Select the Connector with one click. Move the cursor to the entity from which the Connector will be drawn.

births

b) Click and hold. Drag the cursor toward the target.

Population

births

c) When the connector makes contact with the target, the target will highlight/fill in. Release the click to forge the connection. The cursor reverts to the Hand. Click and drag on the circle end of the Connector to move it.

Population

births

A.2.11 Defining Graphs and Tables

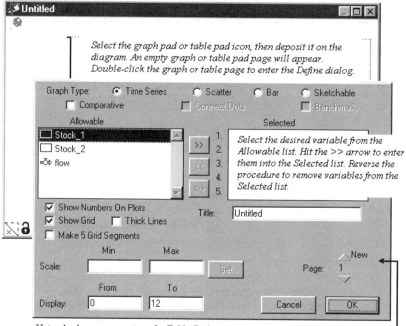

Note: Analogous operations for Table Pads.

Create as many new graph pad pages as you'd like.

A.2.12 Dynamite Operations on Graphs and Tables

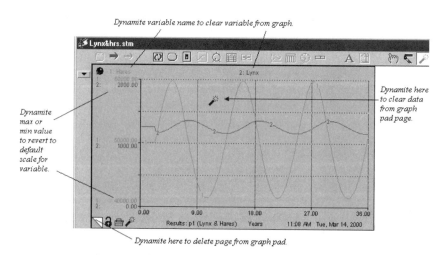

Dynamite variable name to clear variable from graph.

Dynamite here to clear data from graph pad page.

Dynamite max or min value to revert to default scale for variable.

Dynamite here to delete page from graph pad.

Note: Analogous operations for Table Pads.

Index

HIGH PERFORMANCE SYSTEMS, INC., SOFTWARE LICENSE AGREEMENT

Before installing the software, which has been created by HPS and bundled with this book (referred to in this Agreement as "the Software"), please review the following terms and conditions of this Agreement carefully. This is a legal agreement between you and High Performance Systems, Inc. The terms of this Agreement govern your use of the Software. Use of the enclosed software will constitute your acceptance of the terms and conditions of this Agreement.

1. Grant of License.
In consideration of payment of the license fee, which is part of the price you paid for the Software that is bundled with this book, High Performance Systems, Inc., as Licensor, grants to you, as Licensee, a non-exclusive right to use and display this copy of the Software on only one computer (i.e., a single CPU) at only one location at any time. To "use" the Software means that the Software is either loaded in the temporary memory (i.e., RAM) of a computer or installed on the permanent memory of a computer (i.e. hard disk, CD-ROM, etc.). You may use at one time as many copies of the Software for which you have a license. You may install the Software on a common storage device shared by multiple computers, provided that if you have more computers having access to the common storage device than the number of licensed copies of the Software, you must have some software mechanism which locks-out any concurrent users in excess of the number of licensed copies of the Software (an additional license is not needed for the one copy of Software stored on the common storage device accessed by multiple computers).

2. Ownership of Software.
As Licensee, you own the magnetic or other physical media on which the Software is originally or subsequently recorded or fixed, but High Performance Systems, Inc., retains title and ownership of the Software, both as originally recorded and all subsequent copies made of the Software regardless of the form or media in or on which the original or copies may exist. This license is not a sale of the original Software or any copy.

3. Copy Restrictions.
The Software and the accompanying written materials are protected by U.S. Copyright laws. Unauthorized copying of the Software, including Software that has been modified, merged, or included with other software, or of the original written material is expressly forbidden. You may be held legally responsible for any copyright infringement that is caused or encouraged by your failure to abide by the terms of this Agreement. Subject to these restrictions, you may make one (1) copy of the Software solely for back-up purposes provided such back-up copy contains the same proprietary notices as appear in the Software.

4. Use Restrictions.
As the Licensee, you may physically transfer the Software from one computer to another provided that the Software is used on only one computer at a time. You may not distribute copies of the Software to others. You may not modify, adapt, translate, reverse engineer, de-compile, disassemble, or create derivative works based on the Software.

5. Transfer Restrictions.
The Software is licensed to only you, the Licensee, and may not be transferred to anyone else without the prior written consent of High Performance Systems, Inc. Any authorized transferee of the Software shall be bound by the terms and conditions of this Agreement. In no event may you transfer, assign, rent, lease, sell or otherwise dispose of the Software on a temporary or permanent basis except as expressly provided herein.

6. Termination.
This Agreement is effective until terminated. This Agreement will terminate automatically without notice from High Performance Systems, Inc., if you fail to comply with any provision of this Agreement. Upon termination you shall destroy all copies of the Software, including modified copies, if any.

7. Disclaimer of Warranty.
THE SOFTWARE IS PROVIDED "AS IS" WITHOUT WARRANTY OF ANY KIND, EXPRESS OR IMPLIED OF ANY KIND, AND HIGH PERFORMANCE SYSTEMS, INC., SPECIFICALLY DISCLAIMS THE WARRANTIES OF FITNESS FOR A PARTICULAR PURPOSE AND MERCHANTABILITY.

8. Miscellaneous.
This Agreement shall be governed by the laws of the State of New Hampshire and you agree to submit to personal jurisdiction in the State of New Hampshire. This Agreement constitutes the complete and exclusive statement of the terms of the Agreement between you and High Performance Systems, Inc. It supersedes and replaces any previous written or oral agreements and communications relating to the Software. If for any reason a court of competent jurisdiction finds any provision of this Agreement, or portion thereof, to be unenforceable, that provision of the Agreement shall be enforced to the maximum extent permissible so as to effect the intent of the parties, and the remainder of this Agreement shall continue in full force and effect.